W9-BGV-117

THE ANTENNA CONSTRUCTION HANDBOOK FOR HAM, CB & SWL

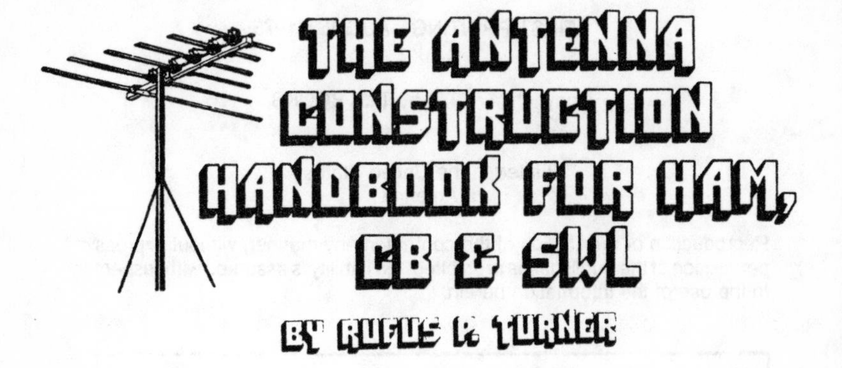
THE ANTENNA
CONSTRUCTION
HANDBOOK FOR HAM,
CB & SWL
BY RUFUS P. TURNER

TAB BOOKS
BLUE RIDGE SUMMIT, PA. 17214

FIRST EDITION

FIRST PRINTING—AUGUST 1978

Printed in the United States
of America

Library of Congress Cataloging in Publication Data

Turner, Rufus P

The antenna construction handbook for ham, CB & SWL.

Includes index.

1. Radio—Antennas—Amateur's manuals. I. Title.

TK9956.T754 621.3841'35 78-15663

ISBN 0-8306-9898-1

ISBN 0-8306-1054-5 pbk.

Preface

The antenna is the last component at a transmitting station and the first at a receiving station. It consequently is a strategic device; and if it does not function efficiently, the finest transmitter or receiver is wasted and the entire communications chain suffers. The antenna is our interface with space.

This book describes, in practical terms, antennas of various types and their applications. The material is intended to serve as a useful guide for those who build their own antennas and also for those who buy them. Its organization focuses on those antennas about which the author and publisher continually receive the most requests for information: ham radio, CB, and SWL types.

Antenna theory can be abstruse. Admittedly, an in-depth understanding of some phases of it demands intimacy with advanced mathematical physics. In this book, however, only essential background theory is presented—in Chapter 1—and this treatment requires no mathematics beyond arithmetic and the most rudimentary algebra. In short, physical explanations in the simplest permissible language are preferred here to mathematical derivations.

We hope that this book may be of service to technicians who work with antennas either vocationally or avocationally.

Rufus P. Turner

Contents

Other TAB books by the author

No. 537 *One Hundred Twenty-Five One-Transistor Projects*
No. 63 *Transistor Circuits*
No. 75 *Transistors: Theory & Practice*
No. 750 *Electronic Conversions, Symbols, & Formulas*
No. 829 *Impedance*
No. 796 *MOSFET Circuits Guidebook—With One Hundred Tested Projects*
No. 887 *106 Easy Electronics Projects–Beyond the Transistor*

Chapter 1
Essential Elementary Theory

This chapter discusses, in simple terms, the theory of electromagnetic waves and of antennas. An understanding of this material is essential to working successfully with antennas in transmission and reception; therefore, it should not be bypassed. Even the experienced technician who may be consulting this book as a refresher might profit from at least a brief examination of this chapter.

The material is presented in the sequence in which it is most often required by the beginning student. But the arrangement is far from sacred, and the reader may move about among the sections as his needs and background dictate. Only the newcomer to antenna theory is cautioned to follow the topics in the order in which they are offered.

We have purposely avoided the higher mathematics and advanced physics which admittedly are important to an in-depth understanding of this subject, since we address the practical man. The ability to handle arithmetic and some algebra and trigonometry will be a sufficient mathematical preparation for our purposes. It is assumed that the reader already knows elementary electricity and that he is experienced in electronics.

ENERGY AND ITS FORMS

Studies and tests show that energy exists throughout the universe, being either at rest (*potential energy*) or in motion (*kinetic energy*). Man is very much concerned with energy and how it can serve him.

No one knows for certain just what energy is, but all of us have a common sense awareness of it. Scientists call it the capacity for doing work, and in comparatively recent times they have demonstrated—what philosophers intuited centuries ago—that energy and matter are interchangeable. All evidence seems to indicate that ultimately there is only one kind of energy, but that it manifests itself in a number of ways. These ways we call *forms of energy*; and although we may not know precisely what energy is, we are directly conscious of these forms and we regularly harness and exploit them. Virtually everything man does or has done to him involves the transforming of energy in one form into energy in another form.

Well-known forms of energy are mechanical, chemical, electrical, radiant, and nuclear. One of these forms—*radiant energy*—includes heat (by the way, all other forms of energy ultimately become heat), light, radio waves, infrared rays, ultraviolet rays, X-rays, and gamma rays, and is the form of energy that travels through space and matter as electromagnetic waves. Radio waves, which chiefly concern electronics people, occupy a very small part of the radiant-energy spectrum, only the region extending from the lowest radio frequencies (10 kHz, or 30,000 meters) to the highest of the superhigh frequencies (30 GHz, or 1 centimeter). And while this spread of approximately 30,000 MHz may at first sight seem enormous, it pales when compared with the total spread of approximately 3×10^{13} MHz of the entire electro-magnetic spectrum from low-frequency alternating currents to gamma rays. It is in the radio-frequency portion of the radiant-energy spectrum that antennas in one form or another are found.

Energy can be stored in an electric field (as in a capacitor) or in a magnetic field (as in an inductor). This important fact underlies the operation of many electronic devices, including antennas. In *LC* circuits and in antennas, for example, electrical energy alternates between the electric field and the magnetic field—i.e., as one field expands, the other collapses, and vice versa—and this action constitutes the phenomenon of oscillation.

NATURE OF WAVES

A *wave* is an undulation of energy whose risings and fallings (or compressions and relaxations) periodically repeat themselves, enabling the effect to travel through a medium such as air, water, or earth, or though empty space. Ocean waves and those created by flipping a rope tied to a tree at the far end are familiar examples easy to observe by eye. Sound waves (in air) and electromagnetic waves (in space) are invisible, but we can detect them through the effects they produce on our hearing (sound) or our instruments (radio). Sound waves consist of alternate condensations and rarefactions of air; radio (electromagnetic) waves are alternate concentrations and rarefactions of energy.

Visible waves, such as those in water, are easily seen to exhibit height and length. Invisible waves likewise have *amplitude* corresponding to height, and *period* corresponding to length and being the time required for the completion of one wave. In mechanical waves, such as those in water and air, particles of the medium are periodically displaced first to one side of their original position of rest, then to the other side. In waves in space, there being no material particles to displace, the amplitude of energy in the wave increases alternately first on one side of zero, then on the other side. Thus, a train of three waves of any sort may be represented by Fig. 1-1. Here, the amplitude (h) may be measured either up from rest (zero), as at A, or down from rest, as at B. And the *wavelength* (λ)—which in mechanical waves is literally the length of one wave, but in waves in free space is the period—may be measured from crest to crest, as at C, or from trough to trough, as at D, or from beginning to end, as at E. Wavelength is expressed in meters, centimeters, or millimeters, and occasionally in feet or inches. Amplitude for visible mechanical waves is expressed in meters, centimeters, feet, or inches; in invisible waves, it is expressed in voltage, current, power, decibels, or loudness, depending upon the nature of the phenomenon. In a

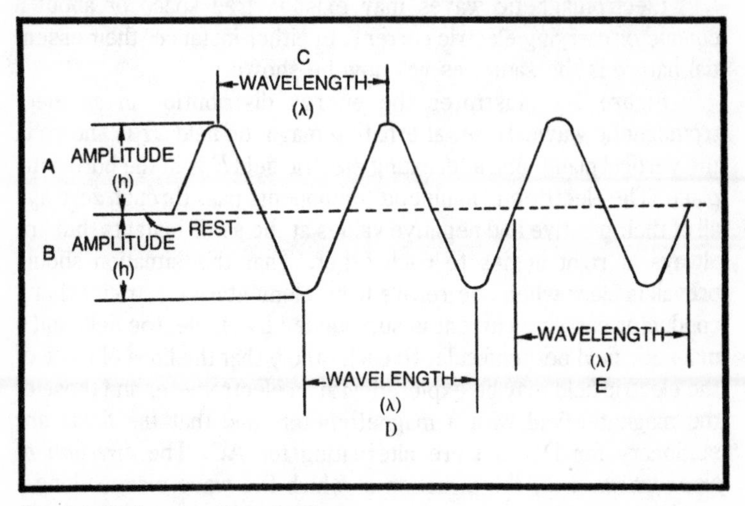

Fig. 1-1. Basic wave characteristics.

wave train, three waves of which are shown in Fig. 1-1, *frequency* is the number of complete waves occurring in 1 second of time. From this description, it is easily seen that the shorter the wavelength, the higher the frequency, and vice versa.

Waves travel through a medium at a velocity which depends upon the kind of wave and the nature of the medium. Thus, sound waves travel through air at approximately 1129 feet per second and through water at approximately 4800 feet per second. Electromagnetic waves (radio, light, heat, etc.) travel through empty space at approximately 186,300 miles per second (300,000,000 meters per second) and more slowly through various materials in proportion to the density of the latter. The velocity of electromagnetic waves in free space is independent of frequency; thus, radio waves and light waves, for example, travel at the same speed. In a dielectric material, the velocity is inversely proportional to the dielectric constant of the material.

ELECTROMAGNETIC WAVES

Electromagnetic waves, as radiant energy, have already been described. At room temperature, all kinds of matter emit some energy in the form of electromagnetic waves (one example is heat), but this emission is very weak. When the temperature of matter is raised, however, or when matter conducts alternating current under the proper conditions, it emits electromagnetic energy of greater strength.

Electromagnetic waves may exist in free space or about a conductor carrying electric current. In either instance, their essential nature is the same, as will now be shown.

Figure 1-2 illustrates the energy distribution in an electromagnetic wave. Here, alternating magnetic field H is shown in the vertical plane, and alternating electric field E is in the horizontal plane. The electric and magnetic components pass through zero and all of their positive and negative values at the same instants, but are always at right angles to each other. That this situation should prevail is clear when one recalls from elementary electricity that a conductor carrying current is surrounded by an electric field and a magnetic field perpendicular to each other, that the lines of force of the electric field may be explored with an electrometer and those of the magnetic field with a magnetometer, and that the fields are stationary for DC and are alternating for AC. The *direction of propagation*—i.e., the direction in which the wave moves along a conductor or through space—is away from the *source*, the point at

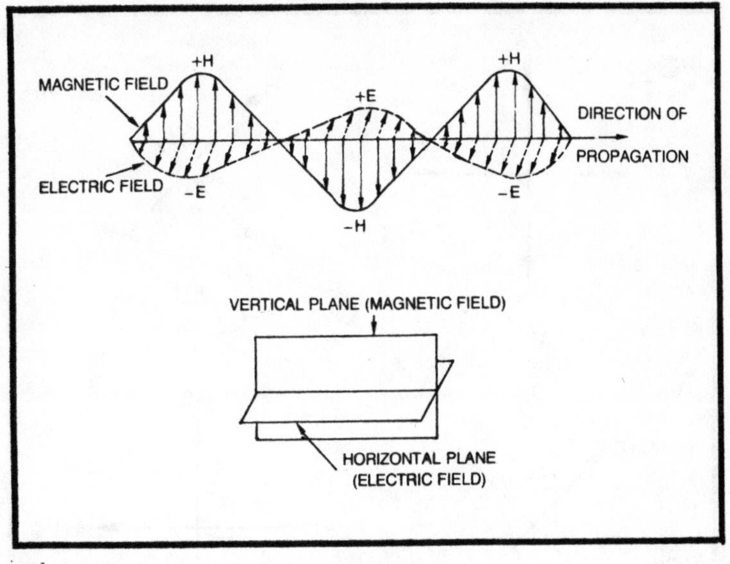

Fig. 1-2. Components of electromagnetic wave.

which the energy originates, and is perpendicular to the motion of both the electric and the magnetic fields, as shown by the horizontal arrow. If the direction of either the electric lines of force or the magnetic lines of force is reversed, the direction of propagation will be reversed.

Although the wave depicted by Fig. 1-2 has its electric component parallel to the surface of the earth, and its magnetic component perpendicular to the surface, this is not the only possibility. The entire figure may be rotated around the line of propagation to any other position (see Fig. 1-3) to show the relationships in a practical wave; however, the magnetic component, electric component, and direction of propagation remain mutually perpendicular. Thus, at *A* in Fig. 1-3, the wave has the same relationship to the surface of the earth as shown earlier in Fig. 1-2; at *B* and *C*, the electric component is at a 45° angle with the surface, rather than parallel; and at *D*, the magnetic component is parallel with the surface and the electric component is perpendicular.

Figure 1-4 shows another way of representing the electric and magnetic components in a wave. Here, the electric lines of force are parallel to the surface of the earth, and the magnetic lines of force are perpendicular to the surface. The plane of this pattern (i.e., the plane of the page) is termed the *wavefront*. And with the lines running in the direction indicated by the arrowheads, this wavefront

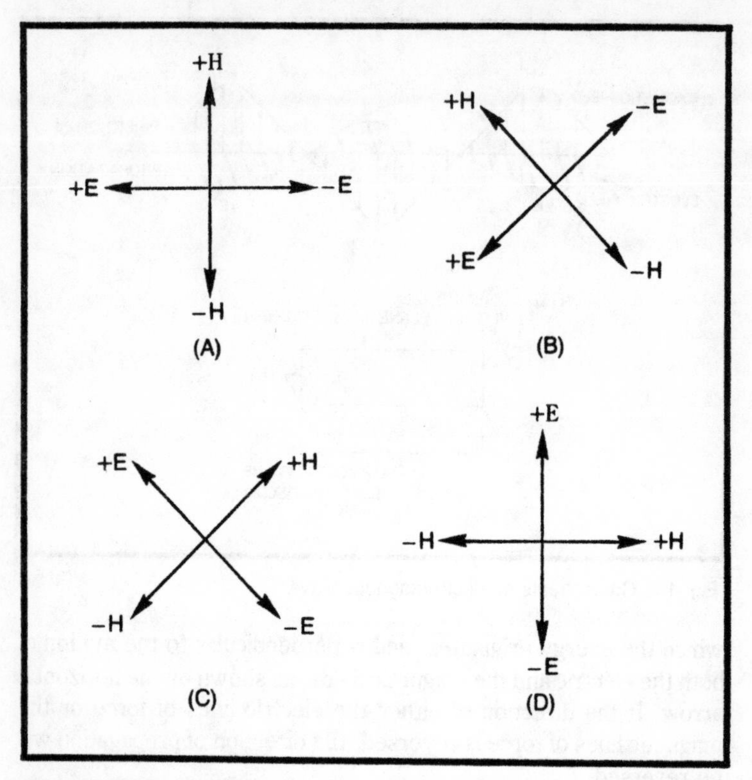

Fig. 1-3. Rotation of wave.

is moving in a direction of propagation into the page. If either the E-lines or the H-lines are reversed, the wavefront will move out from the page.

WAVELENGTH VS FREQUENCY

It has been mentioned earlier that wavelength (λ) is inversely proportional to frequency, and vice versa. This relationship may be expressed:

$$\lambda \; c/f, \text{ and} \qquad (1\text{-}1)$$

$$f \times c/\lambda \qquad (1\text{-}2)$$

where λ = wavelength in meters,
f = frequency in kilohertz, and
c = velocity of light (300,000 km/sec).

Illustrative Example. One of the broadcasts from Standard Frequency Station WWV has a frequency of 5000 kHz. Calculate the corresponding wavelength.

From Equation 1-1,

$$\lambda = 300{,}000/5000 = 60 \text{ meters}$$

Illustrative Example. What frequency corresponds to the amateur radio wavelength of 10 meters?

From Equation 1-2,

$$f = 300{,}000/10 = 30{,}000 \text{ kHz.}$$

Wavelength is not always given in meters, nor frequency in kilohertz, as it is in the above examples. When wavelength is expressed in some other units (such as feet, inches, or centimeters) and frequency in some other units (such as hertz or megahertz), the c term must be suitably multiplied. Tables 1-1 and 1-2 show the resulting formulas for all such calculations. Thus, from Table 1-1, a frequency in megahertz corresponding to a wavelength of 40 meters is $f = 300/40 = 7.5$ MHz; and from Table 1-2, the wavelength in feet corresponding to the frequency of 50 megahertz is $\lambda = 984/50 = 19.7$ feet. These tables do not include frequencies in the gigahertz (10^9 Hz) and terahertz (10^{12} Hz) ranges, since such high frequencies are beyond the scope of the antennas described in this book.

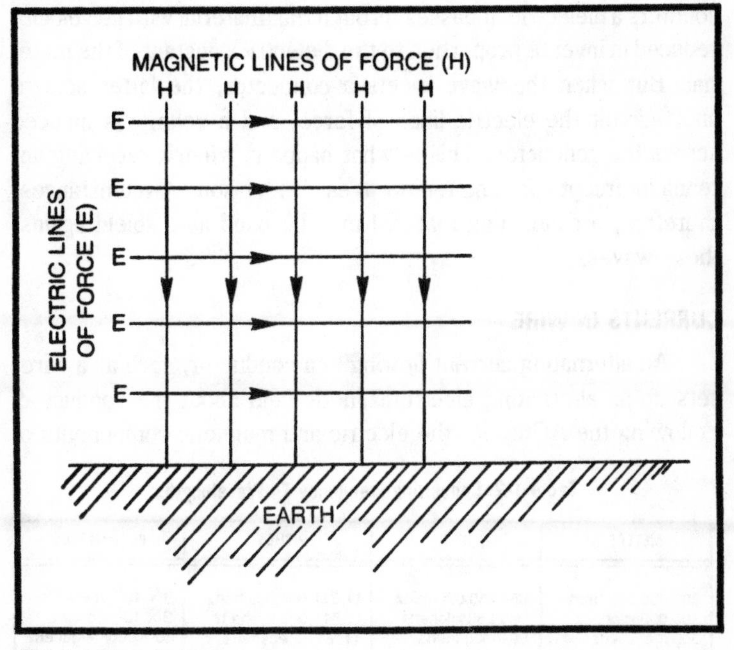

Fig. 1-4. Lines in wave.

Table 1-1. Converting Wavelength to Frequency.

HERTZ	KILOHERTZ	MEGAHERTZ
300,000,000 meters 984,250,000 feet 11,811,000,000 inches 3×10^{10} centimeters	300,000/meters 984,250/feet 11,811,000/inches $3 \times 10^{[illegible]}$/centimeters	300/meters 984/feet 11,811/inches 30,000/centimeters

FIELDS IN SPACE

A sound wave travels from a vibrating source to an ear or other sensor by progressively displacing air particles in its path. An electromagnetic wave, on the contrary, does not need air for its transmission, but travels through space itself (such expressions as "on the air" and "air waves" used in reference to radio arose from early erroneous notions that the atmosphere is the medium through which radio waves travel). The fact that an electromagnetic field can exist in and traverse empty space is clear from the physics classroom demonstration in which waves are passed through a vacuum chamber.

When an electromagnetic wave traveling through space encounters a dielectric, it passes through that material with its velocity reduced in inverse proportion to the dielectric constant of the material. But when the wave meets a conductor, the latter acts to short-circuit the electric lines of force, and a voltage is induced across the conductor. This is what happens when a receiving antenna intercepts passing radio waves. Under some circumstances, therefore, a conducting material may be used as a shield against these waves.

CURRENTS IN WIRE

An alternating current flowing in a conductor, such as a wire, sets up an alternating electromagnetic field about the conductor. Following the AC cycle, the electric and magnetic components of

Table 1-2. Converting Frequency To Wavelength.

METERS	FEET	INCHES	CENTIMETERS
300,000,000/hertz 300,000/kilohertz 300/megahertz	984,250,000/hertz 984,250/kilohertz 984/megahertz	11,811,000,000/hertz 11,811,000/kilohertz 11,811/megahertz	3×10^{10} hertz 3×10^{7} kilohertz 300,000/megahertz

this field expand from zero to maximum intensity, then fall to zero (this corresponding to the positive half-cycle of the current); then they again expand to maximum intensity and fall to zero (this corresponding to the negative half-cycle). Each time the cycle is maximum, the magnetic field is expanded its greatest distance from the wire; and each time the cycle is zero, the magnetic field collapses back into the wire.

As the electrons consitituting the flow of current move along the wire, a rising and falling of electric and magnetic intensity (i.e., a wave) therefore moves along the wire in the direction of electron flow. If the wire is long enough, the losses it introduces will progressively reduce the amplitude of the cycles, as shown in Fig. 1-5, where the wire is 5.5 wavelengths long.

STANDING WAVES

As electrons move along a wire, the magnetic field associated with them accompanies them along the wire. If a pulse is applied to one end of a wire, the electrons move toward the other (open) end of the wire. But when they reach the open end, there is no place for them to go (the open end therefore is equivalent to an infinite impedance) and so the current dies. With the cessation of current, the surrounding magnetic field collapses into the wire, inducing a current which flows in the opposite direction back to the pulse generator. Thus, energy is reflected back along the wire from the open end.

In practice, pulses are continuously applied to a wire and are continuously reflected back from the open end. Under this condition, incident (originated) pulses and reflected (returning) pulses are traveling along the wire at the same time. And, depending upon the frequency of the energy and the length of the wire, the pulses may

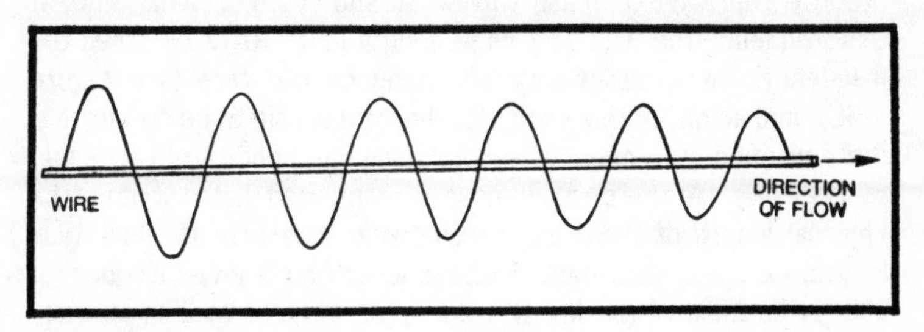

Fig. 1-5. AC in wire.

either reinforce, reduce, or cancel each other at various points along the wire. Therefore, at any point along the wire, the net energy is the algebraic sum of the incident energy and the reflected energy at that point. This produces a series of equally spaced voltage and current points regularly ranging between maximum and zero (or minimum) along the wire. This energy distribution is an interference pattern resulting from the interaction between the incident and reflected components; it does not move along the wire, though the electrons and fields that cause it do themselves move. Because it is stationary, it is called a *standing wave*.

Figure 1-6 shows examples of standing waves. Here, RF generator *GEN* is inserted at the center of a half-wave-long wire. Each end of the wire being open, these ends present infinite impedance to energy flowing along the line. The generator frequency corresponds to 1 wavelength (i.e., $f_{MHz} = 300/\lambda_{meters}$). The algebraic sums of incident and reflected energy at points along the wire give a stationary voltage distribution (dotted line) and stationary current distribution (dashed line), with current maximum at the center of the wire (Fig. 1-6A) and zero at each end, and voltage zero at the center and maximum at each end. The maximum points are termed *loops*, and the zero (or, sometimes, minimum) points *nodes*.

An RF voltmeter will show the various voltage values along the wire, corresponding to the voltage standing-wave pattern. Likewise, if an RF ammeter could be inserted into the wire successively at a number of points, it would show the current values corresponding to the current standing-wave pattern. At any point along the wire, the current and voltage values are different (as one increases, the other decreases), but their product, EI, which indicates power, P, remains constant.

Figure 1-6A shows that when generator frequency f corresponds to 1 wavelength, a single half-wave stands on the half-wave-long wire. With this same length of wire, twice the frequency will produce a full wave (two half-waves), as shown in Fig. 1-6B, since at this frequency the wire is 1 wave-length long. At three times the frequency, three half-waves will stand on the wire (see Figure 1-6C), and so on. In this example, the wire is said to be resonant at the fundamental frequency, second harmonic, and third harmonic, respectively. (At this point, it must be mentioned that in practice the physical length of the wire is somewhat less than the electrical length, i.e., less than the calculated length for a given frequency. This is because of the lower velocity of waves traveling in wire, compared with those traveling in free space. The physical length in

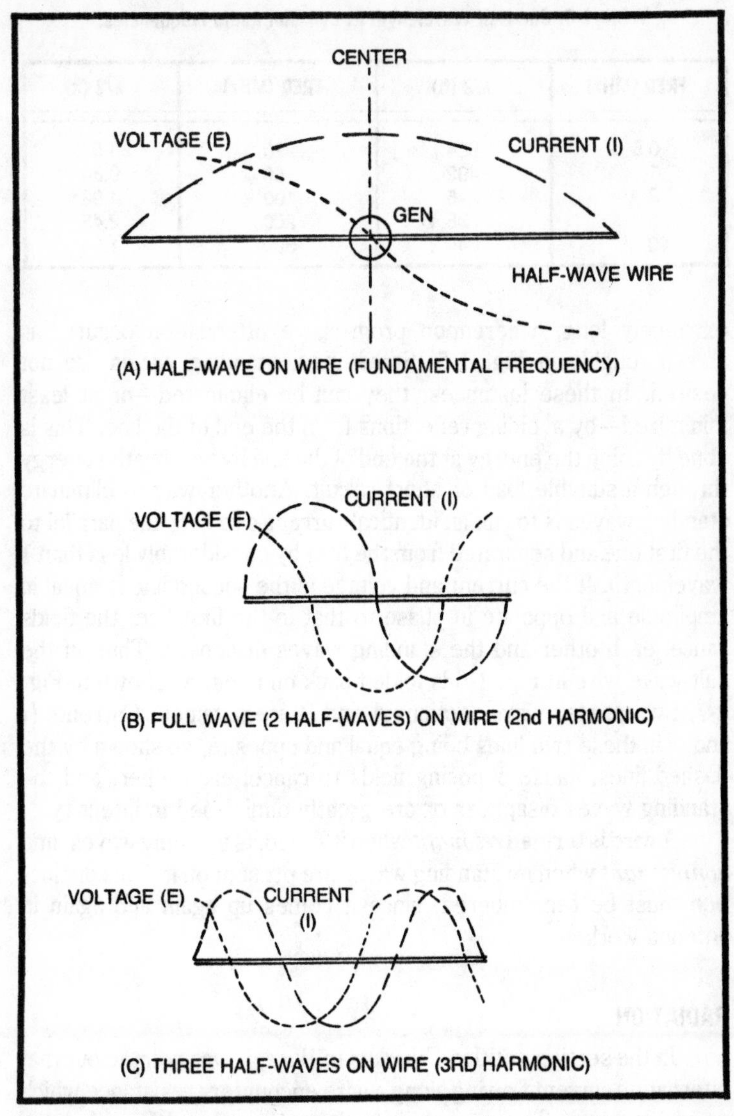

Fig. 1-6. Standing waves.

meters of a half-wave wire is one-half of 0.95 wavelength in meters. Thus, Table 1-3 shows *electrical* lengths which must be multiplied by 0.95 for physical lengths of half-wave wires.)

Whenever there are standing waves, the amplitude of voltage and current varies along a line; when there are no standing waves, the current and voltage are constant at every point unless the line is

Table 1-3. One-Half Wavelength At Common Radio Frequencies.

FREQ (MHz)	λ/2 (ft)	FREQ (MHz)	λ/2 (ft)
0.5	984	20	24.5
1	492	50	9.84
2	245	100	4.92
5	98.4	200	2.45
10	49.2	300	1.54

extremely long, whereupon progressive attenuation occurs, as shown roughly in Fig. 1-5. Sometimes, standing waves are not desired. In these instances, they can be eliminated—or at least minimized—by avoiding reflections from the end of the line. This is done by using the energy at the end of the line by passing the energy through a suitable load or short circuit. Another way to eliminate standing waves is to run an identical current-carrying line parallel to the first one and separated from the first by considerably less than 1 wavelength. If the current and voltage in the second line is equal in amplitude and opposite in phase to that in the first line, the fields cancel each other and the standing waves disappear. Thus, if the half-wave wire in Fig. 1-6 is folded back on itself, as shown in Fig. 1-7, two quarter-wave sections A and B are obtained. Currents I_1 and I_2 in these two lines being equal and opposite, as shown by the dashed lines, cause opposing fields to cancel each other, and the standing waves disappear or are greatly diminished in intensity.

A wire is termed *resonant* when it supports standing waves, and *nonresonant* when no standing waves are present on it. This distinction must be remembered, since it comes up again and again in antenna work.

RADIATION

In the section entitled *Currents in Wire*, it was pointed out that alternating current flowing along a wire encounters resistance which results in losses. The most obvious power loss, $P = I^2R$, is that due to the resistance of the wire. Some power also is lost through ohmic (i.e., in phase) leakage over a path extending from the wire, through supporting insulators, to earth or to a companion wire. A third form of loss is *radiation*. This is the escape of energy from the wire into space, in the form of electromagnetic waves. Radiation, which thus is a loss and therefore ordinarily a defect, is the mechanism that makes radio and TV possible.

Under ordinary circumstances, radiation losses are relatively small. But they increase as the length of the wire becomes a half-wavelength or longer at the frequency of the current, and vice versa. Thus, at the power-line frequency, 60 Hz, radiation losses are infinitesimal because a half-wavelength at 60 Hz is equal to 1553 miles, a very long length for a single uninterrupted power line. But at the higher radio frequency of 10 MHz, a half-wavelength is 49.2 feet, an easily obtained length in a wire, and virtually all of the power is radiated from a wire of that length. The purpose of a transmitting antenna is to radiate power, whereas the purpose of a *transmission line* of any kind is to carry power (as from a transmitter to an antenna) without radiating it. Table 1-3 shows half-wavelengths corresponding to 10 common radio frequencies.

The exact mechanism of radiation is impossible to explain without the aid of advanced mathematics, but a sufficiently clear picture of its nature may be gained through familiarity with the material in the preceding sections of this chapter. For all practical purposes, it is sufficient to understand that as the frequency increases more and more energy leaves a conductor and is radiated into space as electromagnetic waves, that for best results the length of the conductor must be an integral multiple or submultiple of 1 wavelength at the operating frequency, and that standing waves must be present on the conductor. Maximum radiation takes place from current loops on a conductor. The fields move away from a radiator and travel through space at the speed of light: 300,000 kilometers (186,300 miles) per second.

While any metallic body can function to some extent as a radiator, it is convenient to think of a radiator as a straight wire of the proper length, and indeed the radiating part of many antennas con-

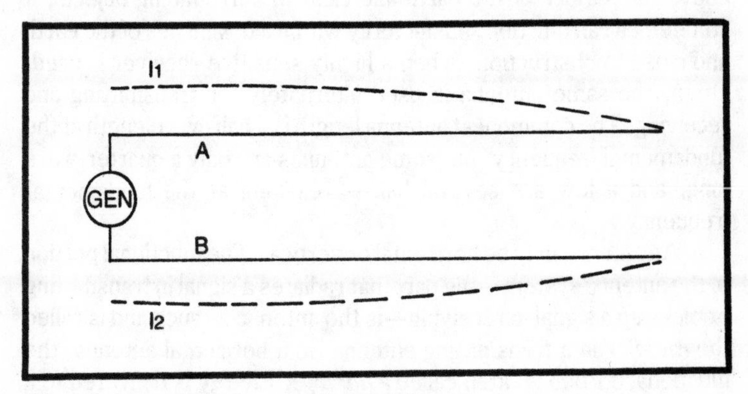

Fig. 1-7. Arrangement for canceling radiation.

sists of just that. It is not imperative, though, that wire be used; many radiators are made of rod or tubing.

The section entitled *Propagation* later in this chapter describes how electromagnetic energy moves from the antenna and through space.

BASIC NATURE AND FUNCTION OF ANTENNAS

Rudimentarily, an antenna consists of a properly dimensioned wire or rod for transmitting or receiving RF energy in the form of radio waves. Energy is supplied to the *transmitting antenna* by a suitable RF generator, such as a radio or TV transmitter. The electromagnetic fields of passing radio waves cut the *receiving antenna*, inducing in it RF voltages which are utilized by the receiving equipment.

The ostensibly simple wire or rod is actually a distributed circuit containing inductance L, capacitance C, and resistance R. In practice, each of these components is relatively small, but becomes more conspicuous as the operating frequency is increased. Therefore, the antenna, however simple, exhibits reactance X_L and X_C and impedance Z. The distributed nature of the equivalent LCR circuit of the antenna results in more effective radiation than could be obtained from an equivalent lumped LCR circuit.

For the antenna to be an efficient radiator, it must be supported well away from surrounding objects and well above the surface of the earth, and it must be well insulated. RF power is supplied to a transmitting antenna through a transmission line, which itself ideally does not radiate. In order for this antenna to radiate, standing waves must be present on it. A receiving antenna likewise must be well insulated. And, though it too benefits from being supported well above the surface of the earth and clear of surrounding objects, it sometimes can function satisfactorily within a few inches of the earth and close to obstructions when a highly sensitive receiver is used. Often, the same antenna is used alternately for transmitting and receiving. The commonest antenna length is a half-wavelength at the fundamental frequency, but some antennas are only a quarter-wave long, and a few are several half-waves long at the fundamental frequency.

An antenna may be horizontal or vertical. The functional portion of the antenna system—the part that radiates a signal in transmitting or picks up a signal in receiving—is the antenna as such and is called the *radiator* in a transmitting antenna (in a horizontal antenna, the functional portion is often called a *flat top*). Energy is delivered to a transmitting antenna by means of a transmission line (sometimes called a *feeder*), and is conducted from a receiving antenna by means of a transmission line, feeder, or *lead-in*.

Ideally, a transmission line, as stated above, does not radiate. This means that no standing waves (or, in practice, only a very small amount of standing-wave energy) may be present on the line. The method of canceling standing waves shown in Fig. 1-7 is often employed for this purpose. That is, the arrangement shown in Fig. 1-7 is used as a two-wire transmission line (pair of feeders) and the antenna is connected to the free ends of wires *A* and *B*. Thus, the feeders may be connected to the center of the antenna (Fig. 1-8A) or to one end of the antenna (Fig. 1-8B).

BASIC ANTENNA TYPES

There are innumerable ways to classify antennas. At the basic level, a simple first breakdown is *vertical* and *horizontal*. Next, either a vertical or a horizontal antenna may be *grounded* or *ungrounded*. The terms vertical and horizontal hardly need explanation, the vertical antenna being erected perpendicular to the surface of the earth, and the horizontal antenna parallel to the surface. These terms are sometimes used when either type actually is erected at an angle. Whether the vertical or the horizontal is chosen depends upon space restrictions, wavelength, desired directional features, and site (whether vehicle, building, open area, etc.). The terms *grounded* and *ungrounded* require further explanation (see below).

Grounded Antenna

With this type, one output terminal of the transmitter or input terminal of the receiver is connected to the antenna, and one terminal is connected to earth. Simple examples are shown in Fig. 1-9 (Figs. 1-9A and 1-9C, vertical; Fig. 1-9B, horizontal). The grounded antenna, also called *Marconi* antenna, is usually a quarter-wave long, but it may be considered to perform somewhat as a half-wave antenna, since a good ground supplies a mirror image of the quarter-wave standing-wave current distribution to complete the half-wave. Thus, in Fig. 1-9C the above-ground quarter-wave is indicated by the solid curve, and the below-ground image by the dashed curve. In the horizontal type (Fig. 1-9B), the total length of the system (horizontal flat top plus the vertical feeder) equals a quarter-wavelength at the fundamental frequency.

The vertical Marconi antenna is very useful where horizontal space is restricted, and the horizontal type is often used when for some reason a more elaborate antenna is impractical. A metal tower

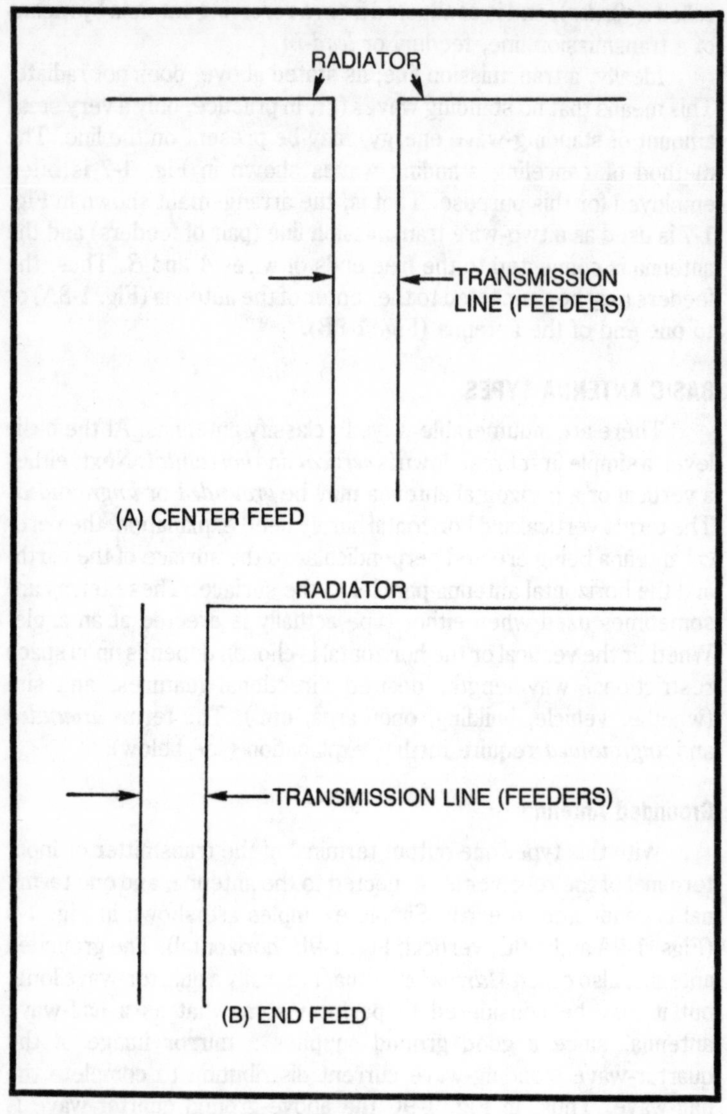

Fig. 1-8. Antenna and transmission line.

sometimes serves as a vertical Marconi antenna. For maximum efficiency, however, the Marconi antenna requires a very-low-resistance ground—something that is hard to obtain, except on the water. In lieu of a ground, a *counterpoise* sometimes is employed. This consists of a number of wires, each a quarter-wavelength or longer, supported just above the ground and insulated from it, and radiating out from a common point to which they are connected together. The counter-poise operates by offering a high capacitance to ground and sometimes gives results equal to those obtained with a low-resistance ground connection, but this arrangement is practical only in a fairly large open area that preferably is fenced in.

Ungrounded Antenna

The faults and difficulties of the grounded antenna are overcome in the *Hertz* antenna. This is an ungrounded antenna having a flat top that is a half-wave long at the fundamental frequency and fed by a transmission line. The basic types of Hertz antennas are shown in Fig. 1-8.

In Fig. 1-8A, the transmission line (feeders) is connected to the center of the half-wave radiator. Because the connection is made at a current loop (voltage node), this type of Hertz antenna is said to be *current fed*. In Fig. 1-8B, the connection at the end of the radiator is made at a voltage loop (current node), so this type of antenna is said to be *voltage fed*. Depending upon the installation and the feeder length, some feeders are tuned (by means of a variable capacitor in one or both feeders), others untuned. Generally, the choice beween center feed and end feed is dictated by the location; that is, whether the radio equipment is at the center or at one end of a permissible half-wave wire.

Hertz antennas commonly have a horizontal flat top and vertical feeders. But the opposite arrangement is sometimes seen, especially in VHF and UHF service. Also, the two halves of the center-fed flat top are sometimes at an angle other than perpendicular to the feeders.

Basic Dipole

The straight-wire, center-fed, half-wave Hertz antenna—termed a *dipole* antenna—is accepted as one of the basic antennas. The performance of other antennas, some of which have been derived from the dipole, is referred to that of the dipole.

ANTENNA AND TRANSMISSION-LINE IMPEDANCE

In general, antennas and transmission lines are linear devices; that is, they consist of straight wires, rods, or tubes that are actually distributed circuits. As such, they present impedance to RF energy. In most instances, the impedance of these devices is resistive.

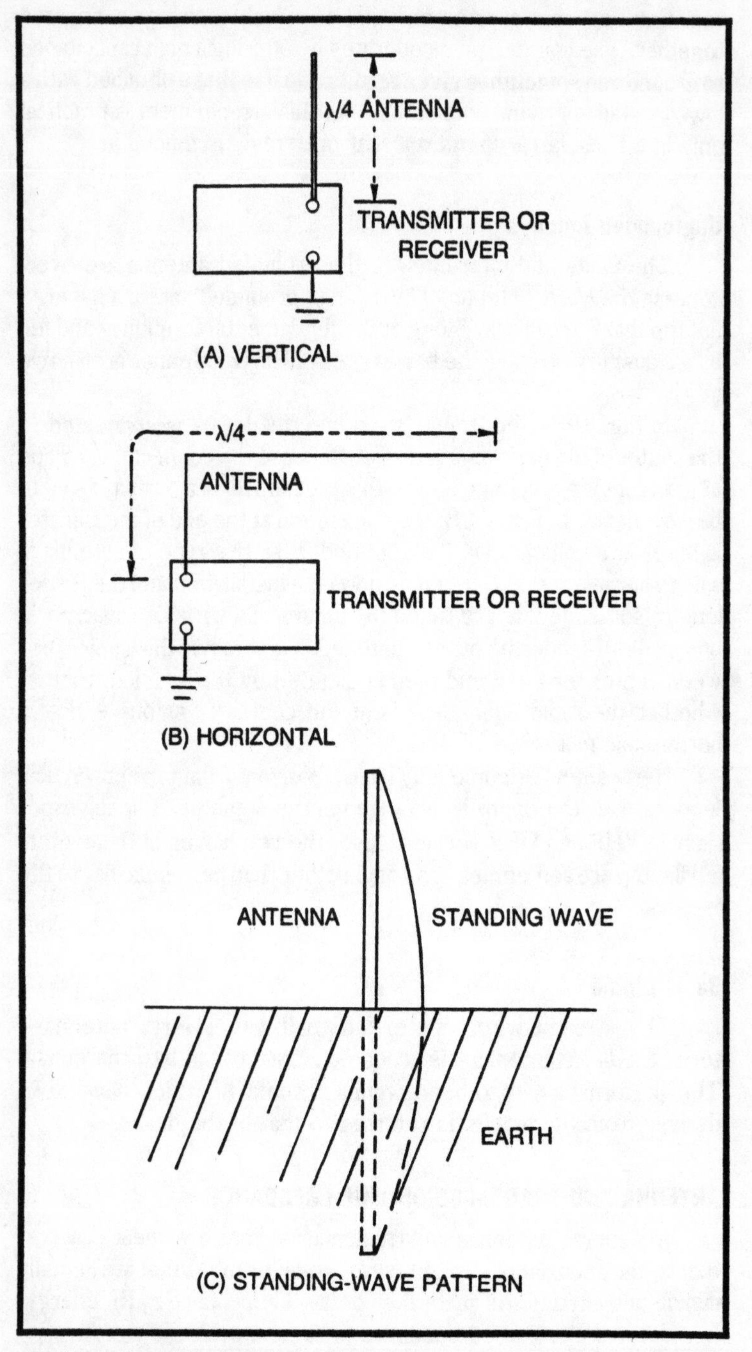

Fig. 1-9. Grounded antenna.

Antennas

An operative antenna is characterized by a pattern of stationary standing waves along its length. This arrangement of loops and nodes constitutes a distribution of current I and voltage E along the length, as shown in Fig. 1-6A for a half-wave antenna operating at its fundamental frequency. Figures 1-6B and 1-6C show the standing-wave pattern when the antenna operates at the second and third harmonics, respectively.

Note in Fig. 1-6A that current is maximum at the center of the wire, rod, or tube, and zero at the ends, whereas voltage is zero at the center and maximum at the ends. At any point along the length of the antenna, the impedance Z is equal to the ratio (E/I) of voltage to current at that particular point. Thus, the impedance is very high at the ends (being theoretically infinite: $Z = E/I = E/0 = \infty$) and is very low at the center (being theoretically zero: $Z = E/I = 0/I = 0$).

A transmitting antenna is visualized as working against an impedance—termed *radiation resistance*—when radiating energy into space. The value of radiation resistance R_R is governed by the height of the antenna above ground. The reason for this is the action of that part of radiated energy which is reflected back from the surface of the earth. This reflected energy arrives at the antenna in or out of phase with energy that is in or leaving the antenna. Depending upon the distance the reflected energy has had to travel to reach the antenna, it either reduces or increases the apparent resistance because of this phase effect. Figure 1-10 shows a plot of theoretical values of radiation resistance at the center of a half-wave antenna for various heights above ground. Observe that the higher the antenna, the more closely R_R approaches a theoretical value of 73.2 ohms. At the ends of the antenna, R_R is several thousand ohms. In practical terms, the radiation resistance is that value of resistance which would, if it were inserted at the center of the antenna, dissipate energy equal to that ordinarily radiated from the antenna. And this is a legitimate concept, for radiated energy is, in effect, energy lost from the antenna.

Transmission Lines

The purpose of a transmission line, as has already been stated, is to conduct RF energy from a transmitter to an antenna, or from an antenna to a receiver, with virtually no radiation from the line. In one of its simplest forms, the transmission line consists of two parallel wires, with the spacing between them small compared with 1

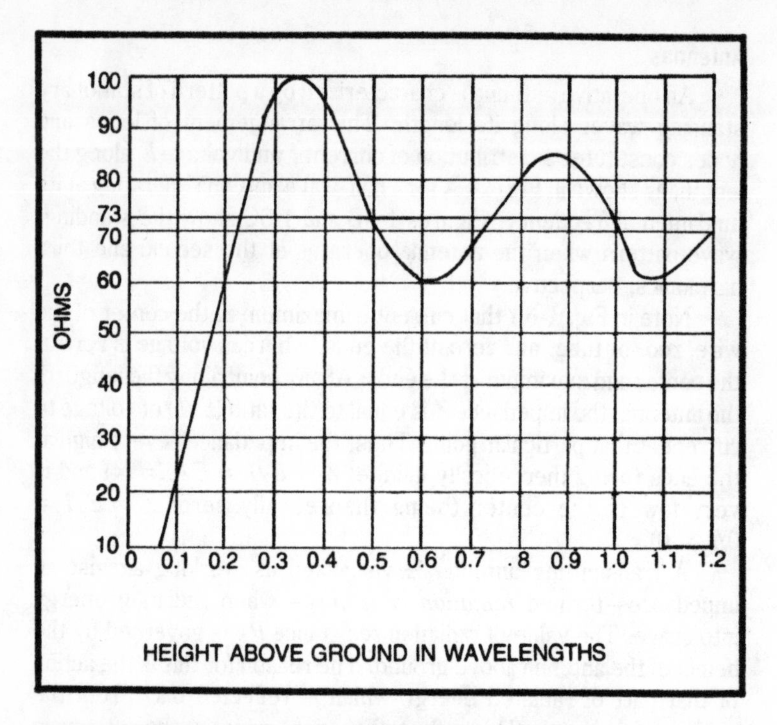

Fig. 1-10. Radiation resistance at center of half-wave antenna.

wavelength. Figure 1-11A shows such a line connected to radio-frequency generator *GEN* at one end and to a load resistor *R* at the other end. Current flows in opposite directions in the two wires, so radiation from the line is effectively canceled. The line has distributed inductance and distributed capacitance, and from these properties the ***characteristic impedance***, Z_O or Z_C, can be calculated (the resistance of the wires is ignored, since it is negligible):

$$Z_O = \sqrt{L/C} \qquad (1\text{-}3)$$

where Z_O is in ohms,

L is in μH per foot, and
C is in pF per foot.

This quantity is termed ***characteristic*** impedance, since for a line of given dimensions the E/I ratio has the same value at any point along the line. It is also called ***surge*** impedance. If the terminating resistance is equal to the characteristic impedance, the resistor absorbs all of the energy, no reflections occur, and therefore no standing waves appear on the line.

Figure 1-11B shows the distribution of current and voltage on an *unterminated* quarter-wave line. From the distribution of current and voltage shown by the dashed and dotted lines, it is clear that various impedances ($Z = E/I$) are available by tapping the line at appropriate opposite points on each wire.

For a two-wire line, Z_0 depends upon the diameter and spacing of the wires:

$$Z_0 = 276 \log_{10} 2S/d \qquad (1\text{-}4)$$

where Z_0 = characteristic impedance in ohms,
S = center-to-center spacing of wires in inches,
d = diameter of wire in inches, and
$\log_{10}$ = common logarithm.

Illustrative Example. The diameter of AWG #12 solid copper wire is 0.081 inch. Calculate the characteristic impedance of a two-wire line consisting of AWG #12 wires spaced 6 inches between centers.

From Equation 1-4,

$$\begin{aligned} Z_0 &= 276 \log_{10} (2 \times 6)/0.081 \\ &= 276 \log_{10} 12/0.081 \\ &= 276 \log_{10} 148.15 \\ &= 276\ (2.1707) = 599.1 \text{ ohms.} \end{aligned}$$

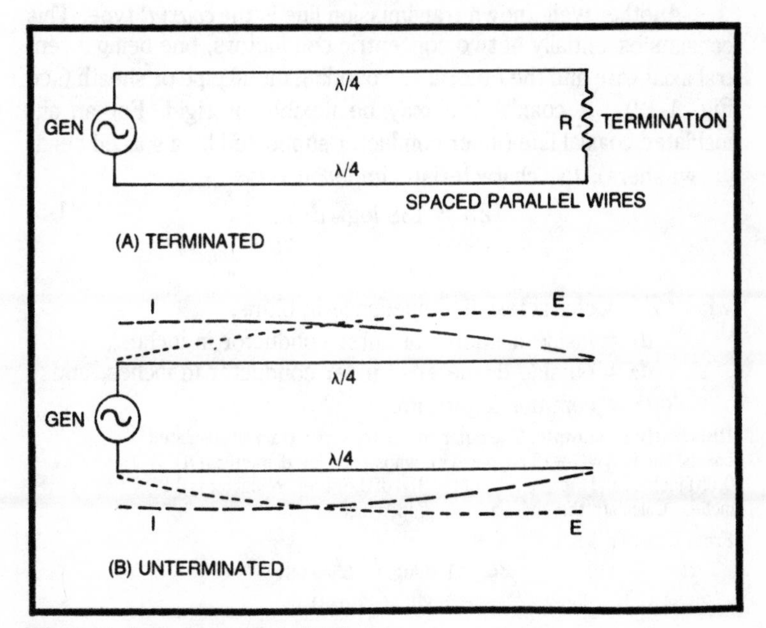

Fig. 1-11. Two-wire transmission line.

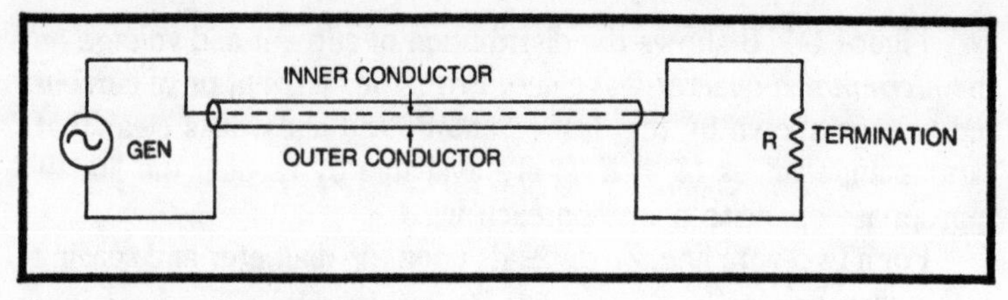

Fig. 1-12. Coaxial transmission line.

NOTE: A pair of AWG #12 wires with 6-inch spacing is commonly called a 600-ohm line.

A closer result (599.78 ohms) is afforded by the equation

$$Z_O = 120 \text{ arc cosh } \frac{[0.5(2S/d)]}{0.5(2S/d)}$$

where Z_O, S, and d are in the same units as in Equation 1-4, and cosh is the hyperbolic cosine.

In its simplest form, the two-wire line is composed of bare wires run in open air. The impedance of an insulated line is somewhat different, however, from that of an open-air line. For example, the 3/8-inch-wide, twin-lead ribbon (two parallel wires molded into a plastic strip) used with TV antennas has an impedance of 300 ohms—about 12 percent higher than the impedance of an open-wire line of the same dimensions.

Another well-known transmission line is the *coaxial* type. This consists essentially of two concentric conductors, one being a central axial wire and the other a surrounding metal pipe or sheath (see Fig. 1-12). A coaxial line may be flexible or rigid. For an air-insulated coaxial line (inner conductor supported by a spaced beads or washers), the characteristic impedance is:

$$Z_O = 138 \log_{10} d_1/d_2 \qquad (1\text{-}5)$$

where Z_O = characteristic impedance in ohms,
d_1 = inside diameter of outer conductor in inches,
d_2 = outside diameter of inner conductor in inches, and
$\log_{10}$ = common logarithm.

Illustrative Example. The inner conductor of a certain air-insulated coaxial line is AWG #12 copper wire whose outside diameter (OD) is 0.081 inches, and the inner diameter (ID) of the outer conductor is 0.25 inches. Calculate the characteristic impedance.

From Equation 1-5,

$$Z_O = 138 \log_{10} 0.25/0.081$$
$$= 138 \log_{10} 3.0864$$
$$= 138(0.489452) = 67.5 \text{ ohms.}$$

When a coaxial line has continuous insulation between outer and inner conductors, the Z_o value obtained with Equation 1-5 must be multiplied by $1/\sqrt{k}$, where k is the dielectric constant of the insulating material. Polyethylene, a common insulation in flexible coaxial lines, has a dielectric constant of 2.3 and therefore requires a multiplier of $1/\sqrt{2.3} = 1/1.516 = 0.659$. Common impedances for commercial polyethylene-insulated coaxial cable are 50, 52, 53.5, 73, and 75 ohms.

PROPAGATION

If a dipole transmitting antenna could be erected in free space, the electromagnetic energy it radiates would be maximum perpendicularly off the center of the antenna and zero off each end in line with the antenna; at points between zero and maximum, the intensity would have intermediate values.

Radiation Pattern

If the field intensity is measured at many points in the horizontal plane of the antenna, the intensity pattern shown in Fig. 1-13 results (here, *B* shows maximum radiation, and *A* and *C* lower radiation at corresponding intermediate points). The two "leaves" of this pattern are termed *lobes*. The antenna radiates not just in the horizontal plane, but also in the vertical plane—above and below a theoretical antenna in free space—and in all intermediate planes between horizontal and vertical. This means that the two-lobe pattern in Fig. 1-13 is found in all of these planes and that in space, the radiation pattern therefore resembles a doughnut with the antenna passing through

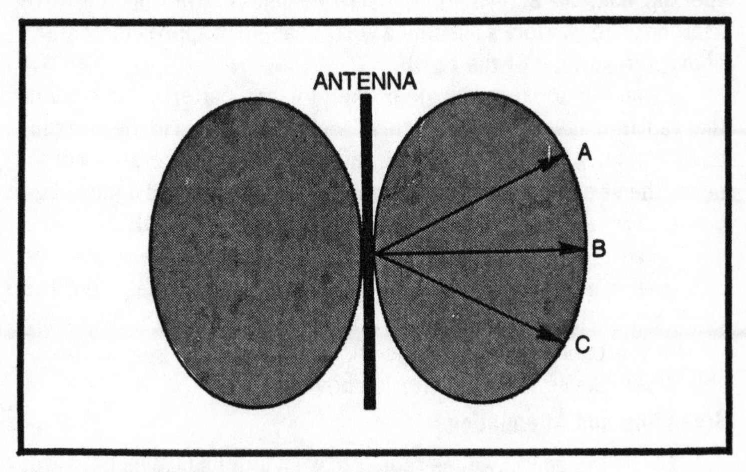

Fig. 1-13. Radiation pattern of ideal dipole antenna.

the tiny hole of the doughnut (a single-plane diagram, such as Fig. 1-13, thus can show only a cross section through the doughnut).

Receivers situated at points perpendicular to the center of the transmitting antenna depicted by Fig. 1-13 pick up the strongest signals, whereas those situated off either end of the transmitting antenna receive either no signal or a very weak one. Conversely, when the dipole is used for receiving, those transmitting stations located to the right, left, above, or below the center of the antenna are received with maximum intensity, whereas those located off either end of the antenna are received weakly or not at all.

Antennas other than the half-wave dipole exhibit radiation patterns having a different number and arrangement of lobes. Figure 1-14 shows the patterns for various selected antennas. With a full-wave antenna (Fig. 1-14A), each of the four lobes extends at an angle, providing four different directions of maximum intensity. Zero (or minimum) intensity is not only off the ends of this antenna, but also off exact center, as well. The 1.5-wave (Fig. 1-14B), also called 3/2-wavelength antenna, provides four major lobes extending at angles from the antenna, two minor lobes perpendicular to the antenna at its center, and six zero (or minimum) points. The 2-wavelength antenna (Fig. 1-14C) provides four major lobes and four minor lobes, each at an angle and in a different direction, and eight zero (or minimum) points. In Figs. 1-14B and 1-14C, the minor lobes are sharper and weaker than the major lobes. Figures 1-14D and 1-14E show patterns for Marconi antennas. For the half-wave ($\lambda/2$) Marconi, the lobes are close to ground, and only half of each shows above the surface (in effect, ground supplies a mirror image of the top half, as shown by the dashed lines). For the full-wave Marconi, the two lobes assume a vertical angle of approximately 30° above the surface of the earth.

From the doughnut shape of the radiation pattern, it is evident that radiation takes place at vertical angles, as well as in the horizontal direction. With some antennas and under certain operating conditions, the vertical-angle radiation may predominate, and it may even be exclusive, as with the full-wave Marconi (Fig. 1-14E).

Since it is impossible to erect an antenna in free space, the radiation characteristic of a practical antenna may differ from the ideal patterns shown in Figs. 1-13 and 1-14, because of the distorting effects of ground and nearby objects. However, it will bear a reasonable resemblance to the patterns shown here.

Spreading and Attenuation

Electromagnetic waves radiated by an antenna travel outward, often in all directions; this is termed *spreading* and is analogous to the

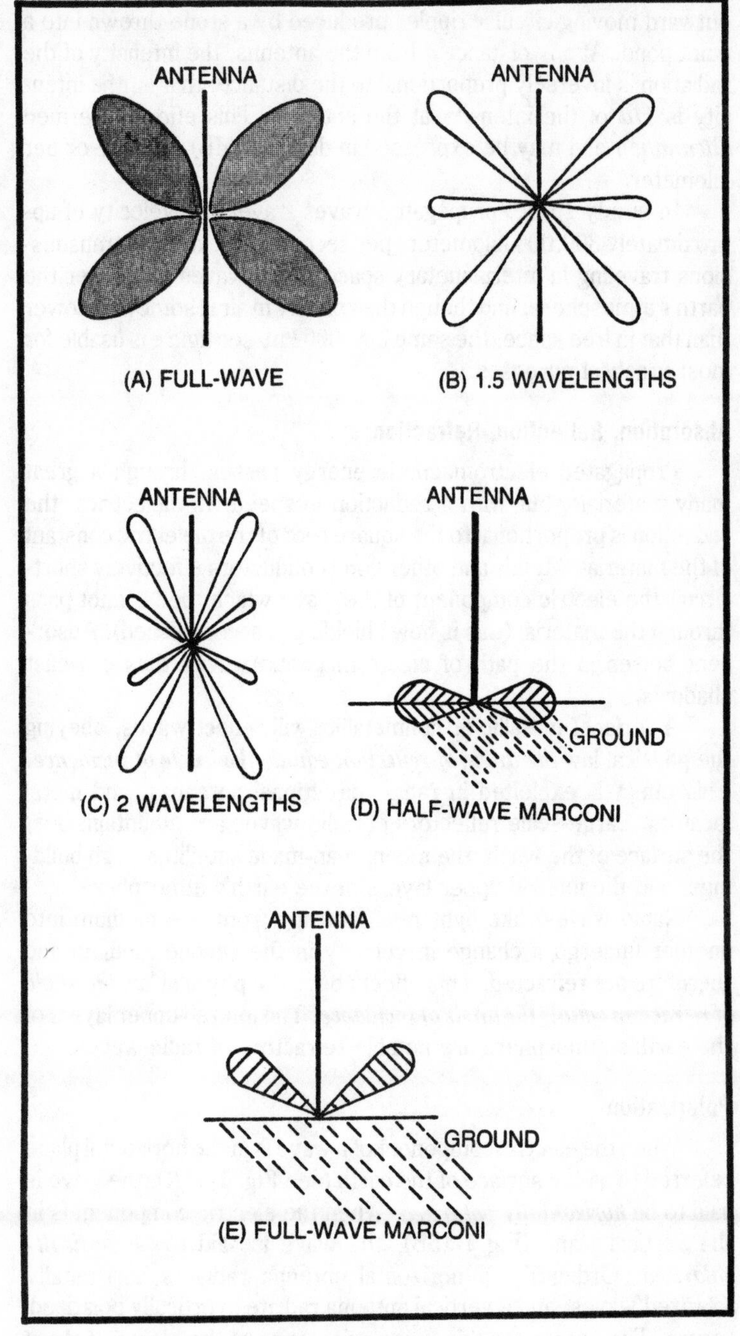

Fig. 1-14. Additional radiation patterns.

outward moving circular ripples produced by a stone thrown into a quiet pond. At any distance *d* from the antenna, the intensity of the radiation is inversely proportional to the distance; that is, the intensity is $1/d$ of the intensity at the antenna. This effect is termed *attenuation* and may be expressed in decibels (dB) per mile or per kilometer.

In empty space, propagated waves travel at a velocity of approximately 300,000 kilometers per second; but except in transmissions traveling in interplanetary space, radio waves encounter the earth's atmosphere, and though the velocity in air is somewhat lower than that in free space, the same 300,000 km/sec figure is usable for most practical purposes.

Absorption, Reflection, Refraction

Propagated electromagnetic energy passes through a great many materials, but with a reduction in speed. In dielectrics, the reduction is proportional to the square root of the dielectric constant of the material. Metals and other good conductors effectively short-circuit the electric component of the wave which then cannot pass through the material (this is how shielding is accomplished). Absorbent bodies in the path of electromagnetic waves thus can cast shadows.

A surface, metallic or nonmetallic, will reflect waves, obeying the physical law *the angle of reflection equals the angle of incidence*. This effect is exploited in radar, directional antennas, and metal locators. Large-scale reflectors of radio waves are mountainsides, the surface of the earth, the moon, man-made satellites, high buildings, and the ionized upper layers of the earth's atmosphere.

Radio waves, like light rays, passing from one medium into another undergo a change in velocity in the second medium and therefore are refracted. This effect obeys the physical law *the angle of refraction equals the angle of incidence*. The ionized upper layers of the earth's atmosphere are notable refractors of radio waves.

Polarization

When the electric component of a wave is in the horizontal plane referred to as the surface of the earth (see Fig. 1-15A), the wave is said to be *horizontally polarized*. When the electric component is in the vertical plane (Fig 1-15B), the wave is said to be *vertically polarized*. Ordinarily, a horizontal antenna radiates horizontally polarized waves, and a vertical antenna radiates vertically polarized waves. The reason for this is the orientation of the electric field of

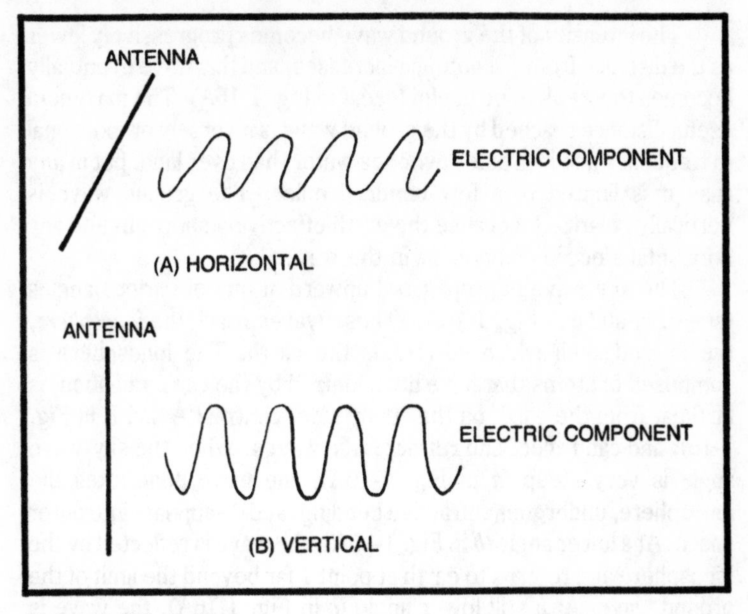

Fig. 1-15. Polarization.

the antenna itself. That is, the electric lines of force about the antenna are in the same plane as the antenna, extending from one end to the other; and since radiation from a straight antenna is broadside, the wave polarization is horizontal for the horizontal antenna, and vertical for the vertical antenna.

Often for best results, it is desirable that transmitting and receiving antennas in the same communication link have identical polarization. For instance, in citizens band communications, limited space makes a vertical antenna mandatory on a car, so the antenna at the home station also is made vertical.

Propagation Components

It has already been shown that an antenna radiates energy in many directions at the same time. Some waves travel in paths along the surface of the earth and constitute the *ground wave*. Because of diffraction, the ground wave follows the curvature of the earth, rather than going off in a straight line at the horizon, but eventually dies out owing to attenuation. Other waves, which travel upward at an angle, are propagated into the upper atmosphere and constitute the *sky wave*. The electric field of a sky wave is at an angle with the surface of the earth; therefore, this wave can be considered to have both horizontal and vertical polarization components.

The intensity of the ground wave becomes progressively lower as the distance from the antenna increases, and this wave eventually becomes too weak to be useful (see *d* in Fig. 1-16A). The maximum useful distance reached by the ground wave is inversely proportional to frequency and is greater over sea water than over land, but in any case it is limited to a few hundred miles. The ground wave is vertically polarized, because the earth effectively short-circuits any horizontal electric component in the wave.

The sky wave is propagated upward at any of various angles (see *a*, *b*, and *c* in Fig. 1-16A). These waves reach the *ionosphere*, the ionized shell which surrounds the earth. The ionosphere is composed of atoms that have been ionized by the sun's radiation, is farthest from the earth on the sunny side (contrast A and B in Fig. 1-16), and can reflect and refract radio waves. When the sky-wave angle is very steep (*a* in Fig. 1-16A), the wave penetrates the ionosphere, undergoes refractive bending, and disappears into outer space. At a lower angle (*b* in Fig. 1-16A), the wave is reflected by the ionosphere and returns to earth at point 1 far beyond the limit of the ground wave. At a still lower angle (*c* in Fig. 1-16A), the wave is reflected at a much lower angle with the inonosphere and returns to earth at a more distant point *2*.No signal is present in the region between the antenna and the point at which the wave returns to earth; this region is termed the *skip zone*. From Fig. 1-16, it is clear that long distances can be covered by means of this skip phenomenon. At certain angles, the sky wave may be reflected several consecutive times by the ionosphere, so that a greater skip distance is covered. This is shown by *a*, *b*, and *c* in Fig. 1-16B where the ionosphere has its lower, nighttime height and the wave is reflected at *a* and again at *b*, returning to earth at the distant point *c*. Sometimes, the earth will reflect a returned wave back to the ionosphere (as from *b'* to *c'* in Fig. 1-16B) from which it is again reflected to a very distant point *d'*. This phenomenon is termed *multihop propagation*.

The skip distance of a sky wave depends upon the angle of radiation, which, in turn, varies with frequency (for a desired communication distance, there are good day frequencies and good night frequencies). It also depends upon position of the sun (i.e., whether transmissions are daytime or nighttime), seasonal variations in the ionosphere, and sunspot activity. The ionosphere is somewhat more complicated than the simple picture given, it being composed of several layers; however, this picture is adequate for the purposes of this book.

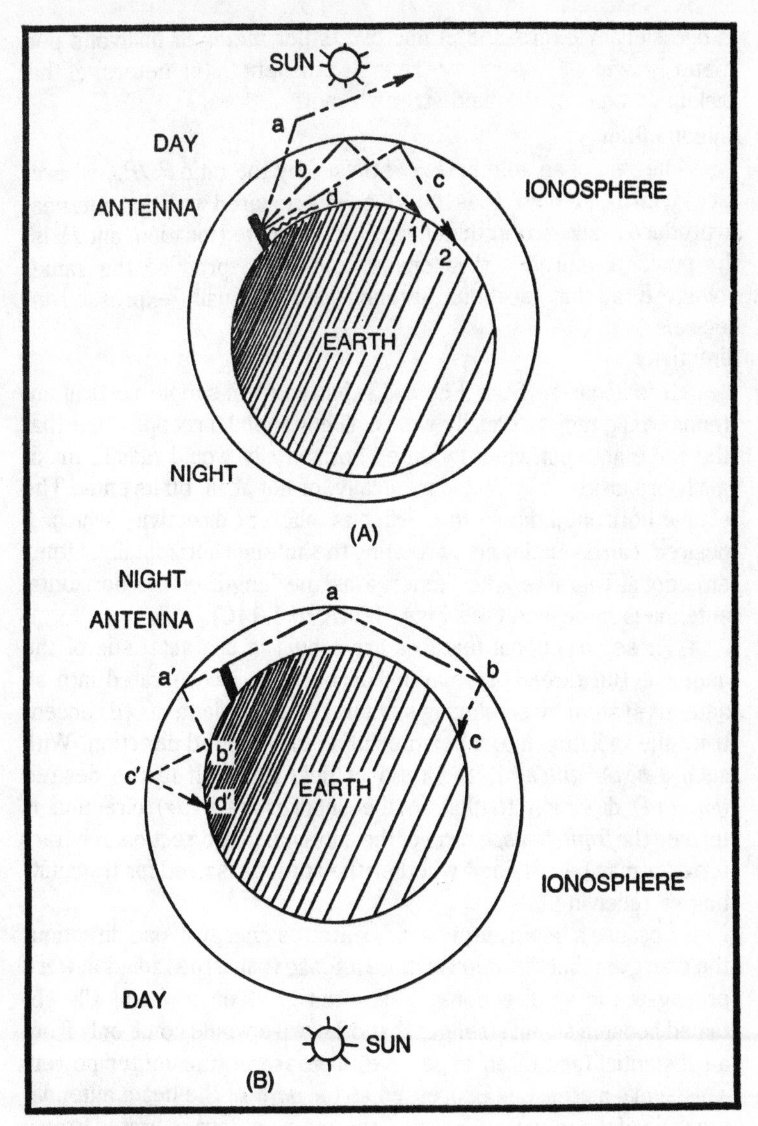

Fig. 1-16. Propagation components.

Field Strength

The field strength of a wave is taken as the intensity of the electric component of the wave at a location far enough away (at least 1 wavelength) to be out of the induction field of the antenna. Field strength is commonly expressed in microvolts or millivolts RMS. It is occasionally expressed in microvolts per meter or millivolts per meter, in which the *per meter* refers to the length (in meters) of the pickup antenna of the field-strength meter.

Antenna Gain

The gain of an antenna is expressed by the ratio P_1/P_2, where (at a given frequency) P_1 is the RF power required with that antenna to produce a field-strength voltage E at a selected location, and P_2 is the power required with a dipole antenna to produce the same voltage E at that location. Antenna gain is usually expressed in decibels.

Directivity

It is apparent from Fig. 1-13 that an ideal simple vertical antenna would radiate equally well in all horizontal directions, but that the same antenna when mounted horizontally would radiate maximally broadside to itself and minimally, or not at all, off its ends. The simple horizontal dipole thus exhibits inherent directivity which, if desired, can be exploited by rotating the antenna horizontally. Other directional characteristics emerge as the length of the horizontal antenna is increased (see Figs. 1-14A to 1-14C).

These directional features are a built-in characteristic of the antenna. But directivity may be deliberately incorporated into an antenna system by employing suitable reflector elements to concentrate the radiation into a beam pointing in a desired direction. With such a *beam antenna*, the ratio of field strength in the desired (*forward*) direction to that in the suppressed (*rear*) direction is termed the *front-to-back ratio* of the antenna. The directional characteristics may be obtained whether the antenna is used for transmitting or receiving.

Because a beam antenna concentrates energy in one direction, the energy in that direction is more intense than if the radiation were propagated in all directions. This often gives an economically obtained boost in signal strength that otherwise would come only from a substantial (and often expensive) increase in transmitter power. This signal increase is expressed as the *gain* of the beam antenna: the ratio of the power radiated by the beam antenna operated from a given transmitter to the power radiated by an equivalent dipole antenna driven by the same transmitter. Thus, the rated 10 dB gain of a three-element, close-spaced beam antenna at 28 MHz means that this beam antenna exhibits a power level 10 times that of an equivalent dipole antenna. (The front-to-back ratio of the three-element antenna is often found to be 30 dB.)

Directional antennas of various types of described in the pertinent chapters which follow.

Chapter 2

Ham Radio Antennas

The transmitting radio amateur has long been in the forefront of antenna development. Always seeking maximum efficiency with his comparatively low-powered equipment (maximum legal input, 1000 watts) and often cramped for antenna space, he has quickly adapted new engineering developments to his particular use and he has been no mean innovator himself.

This chapter describes the principal antennas used in the amateur bands for transmitting and receiving. The material does not go beyond UHF antennas, since only a modicum of ham microwave activity ever has been evidenced. For convenience, however, Table 2-1 lists *all* United States ham-band frequencies, including microwaves allocated for amateur use.

MARCONI ANTENNAS

The Marconi antenna is operated against ground and usually is a quarter-wavelength long. This type finds ham use especially in places where there is insufficient room for a half-wave antenna (a case in point is 160-meter operation, where a half-wave—at 1800 kHz—is 260 feet, but there can be other reasons for its choice).

A Marconi antenna may be eithe rhorizontal or vertical. Figure 2-1 shows the basic types. In each instance, the signal is injected into the antenna at a point as close as possible to ground, since the current loop of the standing wave is at that point (see pattern in Fig. 1-9C, Chapter 1). Also, this reduces the length of the lead from transmitter to ground, a desirable move, as this length is a part of the

Table 2-1. Ham Bands.

BAND	FREQUENCY LIMITS	BAND	FREQUENCY LIMITS
160m	1800–2000 kHz	1.25m	220–225 MHz
80m	3500–4000 kHz	70 cm	420–450 MHz
40m	7000–7300 kHz	–	1215–1300 MHz
20m	14,000–14,350 kHz	–	2300–2450 MHz
15m	21.00–21.45 MHz	–	3300–3500 MHz
10m	28.0–29.7 MHz	–	5650–5925 MHz
6m	50.0–54.0 MHz	–	10,000–10,500 MHz
2m	144–148 MHz	–	24,000–24,050 MHz

quarter wavelength of the antenna. Note that in the horizontal types (Figs. 2-1B and 2-1C), the quarter-wavelength includes both the flat top and the lead-in (since the vertical portion is a part of the quarter-wave radiator, it is not a true *feeder*). In Fig. 2-1C, the flat top is divided into two equal-length sections—*ab* and *ac*—which are electrically in parallel with each other; thus, from ground to ***a*** to ***c*** there is one quarter-wave; and from ground to ***a*** to ***b***, there is another quarter-wave in parallel with the first one.

Figure 2-2 shows rudimentary constructional details of antennas of these types. In Fig. 2-2A, a self-supporting vertical rod or pipe is mounted directly in the ground; in Fig. 2-2B, a self-supporting rod or pipe is insulated from ground by means of a base insulator; in Fig. 2-2C, a vertical rod, pipe, or wire is attached to a wooden mast or pole by means of standoff insulators. Figures 2-2D and 2-2E show common horizontal, wire-type Marconis.

For all types, the length of a quarter-wave of conductor is expressed as:

$$l = 234/f \tag{2-1}$$

where l = length of conductor in feet, and
f = operating frequency in megahertz.

Illustrative Example. What is the length in feet of a quarter-wave antenna for 3500 kHz?

Here, 3500 kHz = 3.5 MHz. From Equation 2-1,
$l = 234/3.5 = 66.9$ feet = 66 feet 11 inches.

Table 2-2 gives the length in feet of quarter-wave Marconi antennas for the 160-, 80-, 40-, 20-, 15-, and 10-meter ham bands. In this table, column 2 shows the high, low, and center frequencies in each band, and column 3 shows the corresponding antenna lengths.

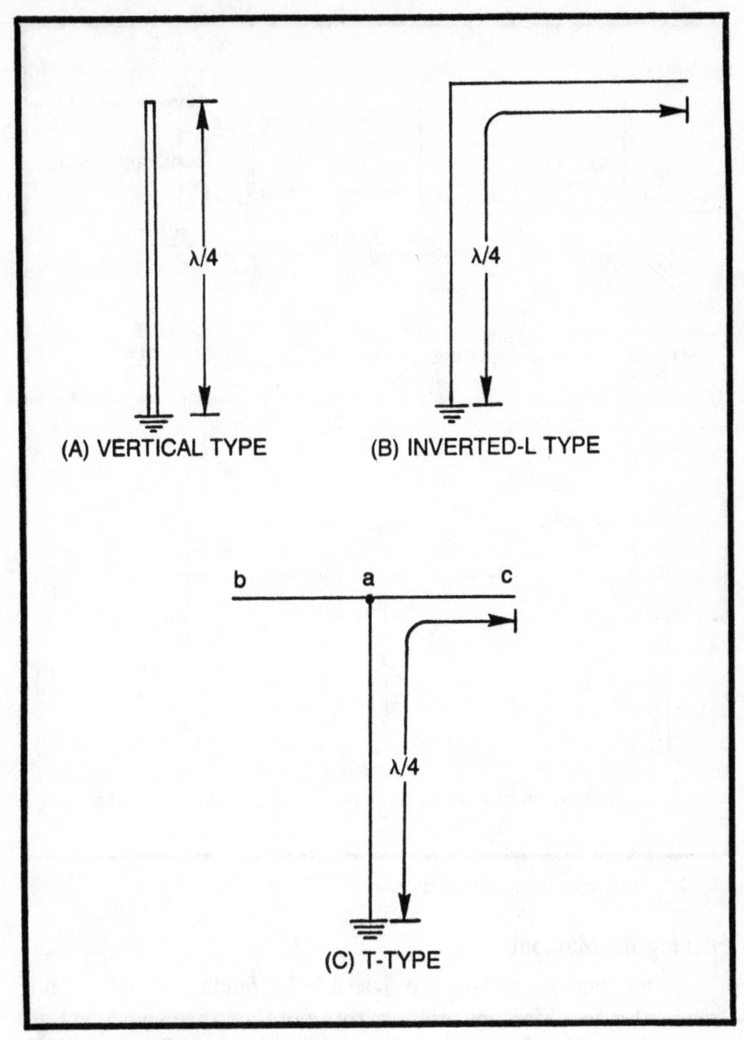

Fig. 2-1. Basic Marconi antennas.

It can be argued that quarter-wave Marconi antennas are impractical at wavelengths shorter than 20 meters, since the required lengths are too short for the antenna to be supported very high above ground; nevertheless, the 15- and 10-meter specifications are included here for the extraordinary case. Certainly at 21 MHz and higher, Hertz antennas are more desirable than the Marconi.

For best performance, every Marconi antenna *must* have a good ground. Pay particular attention, therefore, to the section titled *Ground Connection for Marconi Antenna*.

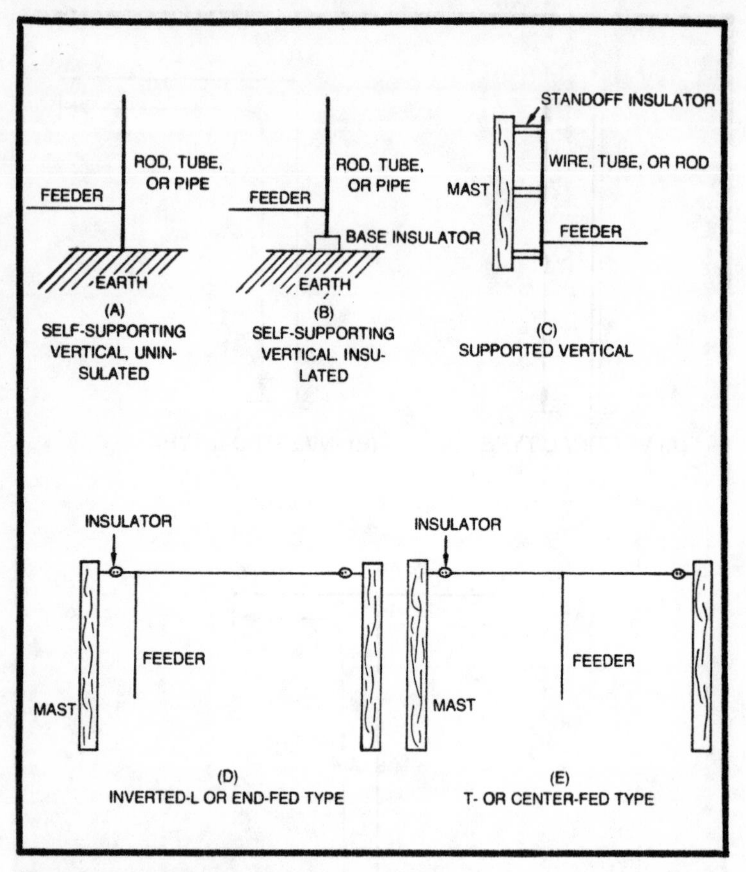

Fig. 2-2. Marconi constructional details.

Feeding the Marconi

A number of systems are available for feeding energy from a transmitter to a Marconi antenna; three of these are shown in Fig. 2-3. In Fig. 2-3A, a 50-ohm coaxial cable (such as type RG8 or RG58) is connected to a low-impedance output-coupling coil at the transmitter end and to the base of an ungrounded vertical (same type as in Fig. 2-2B) at the other end. In Fig. 2-3B, a single-wire line from the transmitter is connected to a grounded vertical (same type as in Fig. 2-2A) at the matched-impedance point (a distance x usually approximately 33 percent of antenna height up from the ground end) at which antenna current is maximum. In Fig. 2-3C, a series-resonant coupler (C1-L2) tuned to resonance at the transmitter frequency is link coupled through L1 to the transmitter and supplies energy to either a horizontal or vertical Marconi.

Peculiarities of Marconi Antennas

1. The Marconi antenna must be mounted as high as possible above ground.
2. The Marconi must be provided with the best available ground.
3. For long-distance communication, this antenna is inferior to the Hertz.
4. Because of losses in the ground connection, the radiation efficiency of the Marconi is inferior to that of the Hertz. (The I^2R losses in the ground circuit may be reduced by minimizing the antenna current by increasing the radiation resistance. One way to do this is to make the antenna length slightly longer than a quarter-wavelength, then to shorten it electrically to a quarter-wave-length by means of series-capacitance tuning.
5. Just as a Marconi antenna that is physically longer than a quarter-wavelength can be shortened to an electrical

Table 2-2. Marconi Antenna Dimensions.

METERS	MEGAHERTZ	λ/4 ANTENNA
160	1.8 1.9 2.0	130'0" 123'2" 117'0"
80	3.5 3.75 4.0	67'10" 62'5" 58'6"
40	7.0 7.15 7.3	33'6" 32'9" 32'0"
20	14.0 14.175 14.35	16'8" 16'6" 16'4"
15	21.0 21.225 21.45	11'2" 11'0" 10'11"
10	28.0 28.85 29.7	8' 4.25" 8'1.25" 7' 10.5"

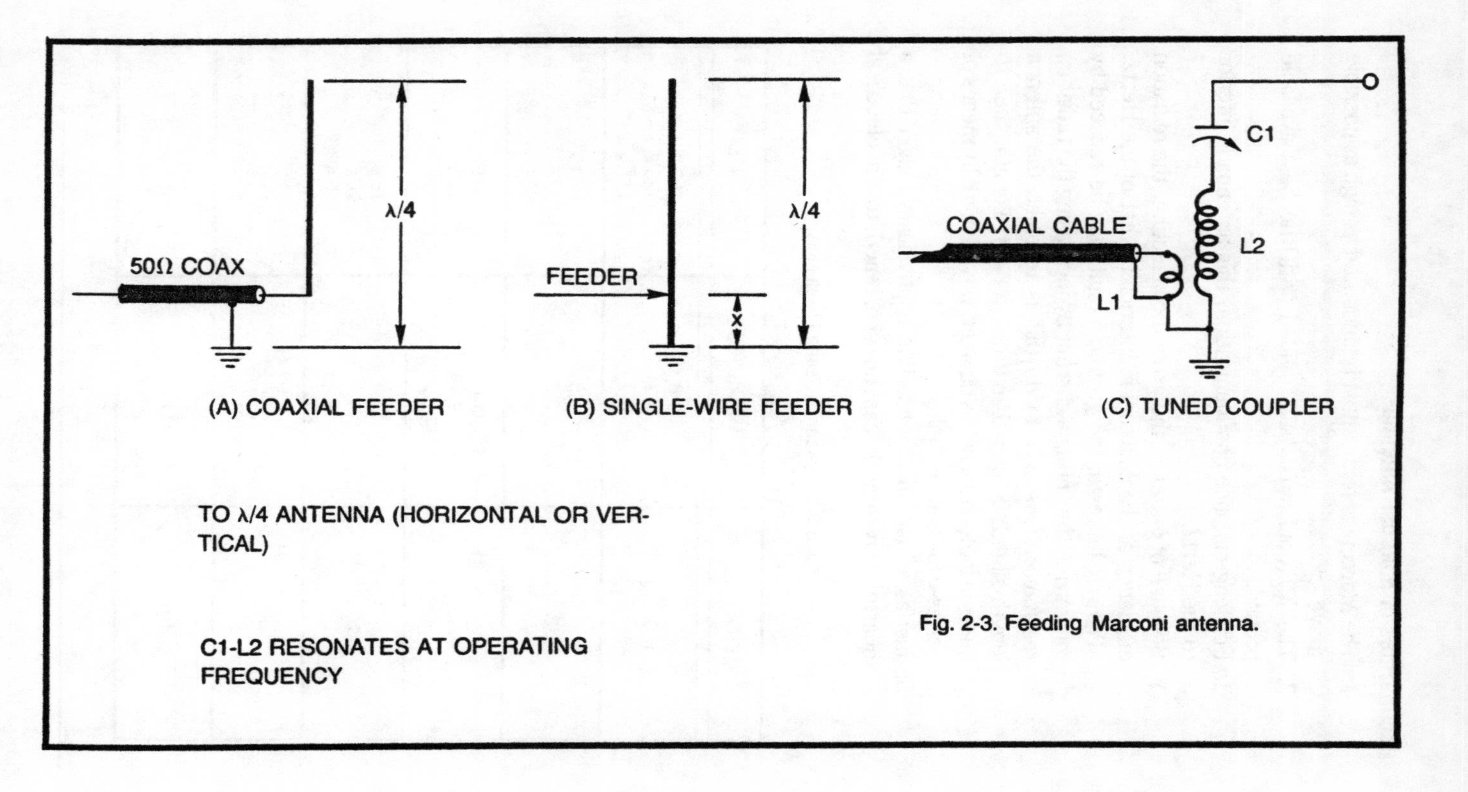

Fig. 2-3. Feeding Marconi antenna.

quarter-wave by means of a series capacitor, an antenna that is physically shorter than a quarter-wavelength can be lengthened to an electrical quarter-wave by means of a series loading coil.

GROUND CONNECTION FOR MARCONI ANTENNAS

A good ground is absolutely essential to efficient operation of a Marconi antenna. The best ground is a connection to a body of water (preferably salt), but this is possible only when the antenna is on board a vessel or is erected on the water's edge. The next best ground is very-low-resistance (usually moist) earth. Most earth, however, is not low resistance, so the best connection must be worked out for a particular location. Several schemes are described in the following paragraphs.

Buried Radials

The best nonmarine type of ground connection consists of a number of radials of bar copper or aluminum wire, each at least a quarter-wave long, soldered, welded, or bonded at a common point (see Fig. 2-4A), and buried in the earth. The wire size should be AWG #10 or #12, and as many radials as possible should be employed. A narrow trough or slot may be dug to receive each horizontal radial, and the earth repacked solidly on top of them after burial. The ground lead connected to the radials should be AWG #10 or #12 wire. The best location for the buried-radial system is directly under the antenna.

Driven Pipes or Rods

A good ground connection can be obtained with several long metal pipes or rods 1 inch or more in diameter driven vertically into the ground as deep as possible and connected together at their tops with an AWG #10 or #12 wire (see Fig. 2-4B). The largest convenient number of pipes should be used, and they should be spread out over an area, rather than being arranged in a single straight line. The best place for the pipe network is directly under the antenna.

When it is impracticable to use a group of pipes, a single pipe or rod sometimes will provide a good ground connection when the diameter of the pipe is not under 1 inch and the pipe is driven deep into the ground (see Fig. 2-4C). Nevertheless the resistance of this type of ground connection must necessarily be higher than that of a group of pipes.

Water Pipe

Since a cold-water pipe usually runs for some distance underground, it can provide a ground connection whose effectiveness depends upon the nature of the soil the pipe touches. For best results, a solid connection must be made to the pipe as close as practicable to the surface of the earth. Because cold-water pipes are present in most buildings, they immediately come to mind for this use. A shortcoming of the pipe, however, is the fact that several sections of pipe are often threaded together, and the plumbing dope used in the couplings tends to insulate one section from another, or at least to introduce undesired resistance between them. It must be noted also that when a connection is made to an extremity of the pipe—as on one of the upper floors of a building—the entire length of pipe between that point and the surface of the earth becomes part of the quarter-wavelength of the antenna system.

Counterpoise

In some places, a reasonably low-resistance ground cannot be obtained. This is particularly true where the soil is dry, rocky, or sandy. An alternative then is to mount the radials described previously under *Buried Radials* horizontally above the surface of the earth, close to the surface but well insulated from it. Such an arrangement is called a *counterpoise* and it operates by providing a high capacitance to ground over a useful area of the surface. For the highest possible capacitance, a large number of radials must be used; the more numerous the radials, the more nearly the counterpoise approximates a large sheet of metal. The counterpoise must be close enough to the surface to provide high capacitance, yet physically high enough to prevent trouble caused by flooding, trash trapping, and similar annoyances. Depending upon local conditions, the height could be anywhere from a few inches to several feet.

A mistaken practice among some hams is to run a single, horizontal, insulated-from-ground, quarter-wave wire under a horizontal antenna or straight away from a vertical antenna and to suppose this to be a one-radial counterpoise. But this single wire cannot provide sufficient capacitance for good counterpoise action, and it tends to function poorly instead as the missing half which would convert the Marconi antenna into a half-wave dipole.

SIMPLE DIPOLE

A simple half-wave dipole fed at the center by a coaxial line is one of the most convenient Hertz-type ham antennas. It is easily

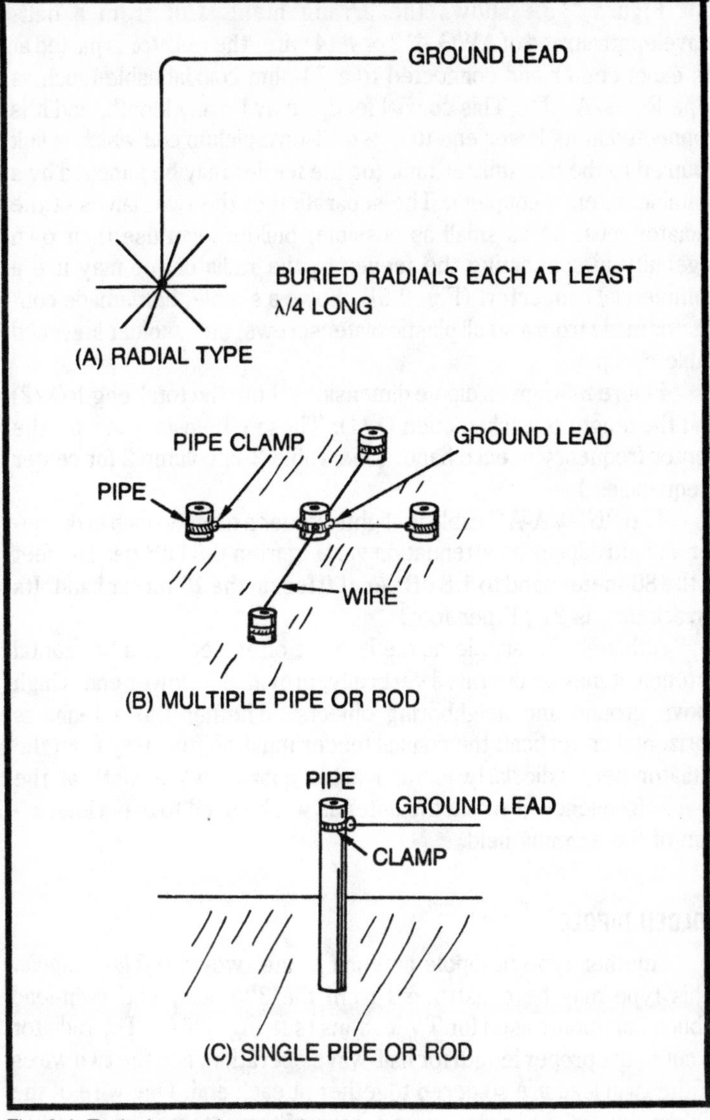

Fig. 2-4. Typical ground connections.

assembled and erected and may be cut for any of the ham bands; however, it is used most often in the 10-to-160-meter bands, other types being preferred at the shorter wavelengths. The simple dipole may be operated at harmonics, as well as at its fundamental frequency, but with somewhat different radiation features. This antenna is also called a simple *doublet*.

Figure 2-5A shows the arrangement. Cut from a half-wavelength strand of AWG #12 or #14 wire, the radiator is parted at its exact center and connected to a 73-ohm coaxial cable (such as type RG59/A-AU). This coaxial feeder may be any length, and it is connected at its lower end to a 2- or 3-turn pickup coil which is link coupled to the transmitter tank (or the feeder may be supplied by a suitable antenna coupler). The separation of the two halves of the radiator must be as small as possible; builders can use their own ingenuity in connecting the feeder to the radiator, or may use a commercial connector. (Fig. 2-5B shows a simple, homemade connector made from a small plastic plate, screws, nuts, solder lugs, and cable clamp.)

Figure 2-5C gives dipole dimensions, both the total length ($\lambda/2$) and the length of each section ($\lambda/4$). These dimensions are for the center frequency in each band. (See Table 2-2, column 2 for center frequencies.)

The RG59/A-AU cable is slightly under a quarter-inch in diameter and introduces an attenuation varying from 0.64 dB per 100 feet in the 80-meter band to 1.8 dB per 100 feet in the 10-meter band. Its capacitance is 21 pF per foot.

Although the simple dipole is most often seen as a horizontal antenna, it may be operated vertically, provided its lower end is high above ground and neighboring objects. Whether the antenna is horizontal or vertical, the coaxial feeder must be run away from the radiator perpendicularly for at least a quarter-wavelength at the lowest frequency at which the antenna will be used to avoid interaction of the antenna field.

FOLDED DIPOLE

Another type of dipole antenna is the two-wire *folded dipole*. This type may be constructed from the 300-ohm, flat, twin-lead ribbon commonly used for TV lead-ins (see Fig. 2-6A). The radiator is cut to the proper length for half-wave operation, and the two wires of the twin lead are soldered together at each end. One wire of the ribbon then is cut at the exact center of the antenna, and another section of twin lead which serves as the feeder is connected to this wire (again see Fig. 2-6A).

Because of the plastic material enclosing the wires, the velocity factor of this dielectric must be taken into account in the calculation of the physical length of the half-wave:

$$l = 403.4/f \qquad (2\text{-}2)$$

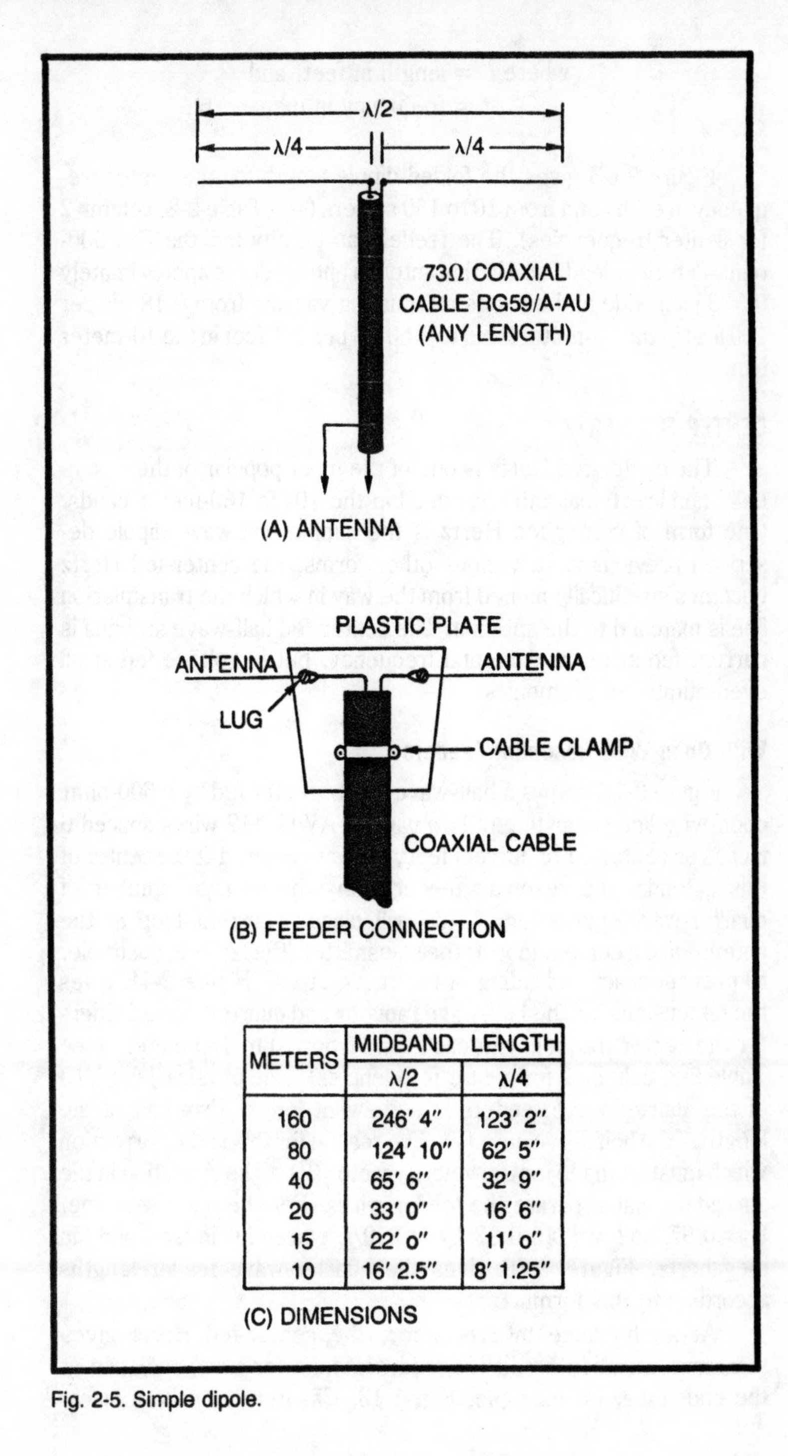

METERS	MIDBAND LENGTH	
	λ/2	λ/4
160	246′ 4″	123′ 2″
80	124′ 10″	62′ 5″
40	65′ 6″	32′ 9″
20	33′ 0″	16′ 6″
15	22′ o″	11′ o″
10	16′ 2.5″	8′ 1.25″

Fig. 2-5. Simple dipole.

where l = length in feet, and
f = frequency in megahertz.

Figure 2-6B gives the folded-dipole length for the center frequency in each band from 10 to 160 meters (see Table 2-2, column 2 for center frequencies). The feeder can be any length. The 300-ohm, flat, twin-lead used in this antenna and feeder is approximately 0.375 inch wide and exhibits attenuation varying from 0.18 dB per 100 feet in the 80-meter band to 0.60 dB per 100 feet in the 10-meter band.

CENTER-FED HERTZ

The center-fed Hertz is one of the most popular of the essentially nondirectional antennas used in the 10- to 160-meter bands. One form of center-fed Hertz is the simple half-wave dipole described previously. In various other forms, the center-fed Hertz becomes specifically named from the way in which the transmission line is matched to the antenna. The center-fed half-wave antenna is current fed at its fundamental frequency, but is voltage fed at all even-numbered harmonics.

With Open-Wire Resonant Feeders

Figure 2-7A shows a half-wave Hertz center fed by a 600-ohm open-wire line consisting of two parallel AWG #12 wires spaced 6 inches on centers. For current feed, such as required at the center of this antenna, the resonant feeders must be an *even* number of quarter-wavelengths long. This will place a current loop at the antenna and a current loop at the transmitter (the latter is desirable, to prevent reactive loading of the transmitter). Figure 2-7B gives the dimensions for the half-wave radiator and quarter-wave feeders for the center frequency in each band from 10 to 160 meters (see Table 2-2, column 2 for center frequencies). The physical length (l) of the quarter-wave feeders is somewhat longer than half of the length of the half-wave antenna. The reason for this is the correction which must be made for the velocity factor (V) of the insulation in the spreaders that separate the feeder wires. For the open-wire line, $V = 0.97$, so $l = [246(0.97)]/f = 239/f$, where l is in feet, and f in megahertz. Figure 2-7B gives the quarter-wave feeder lengths according to this formula.

At its fundamental frequency, the center-fed Hertz gives maximum radiation broadside, and minimum (or zero) radiation at the ends (see, for example, Fig. 1-13, Chapter 1); at harmonics,

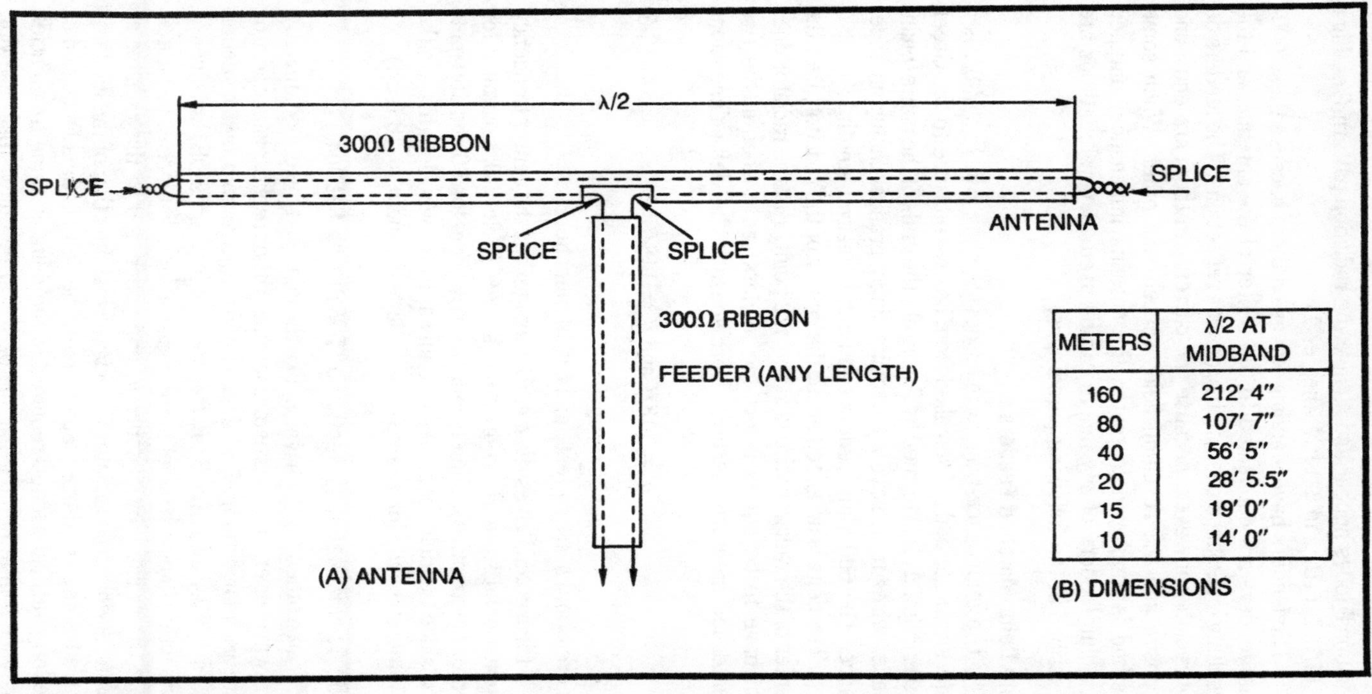

METERS	λ/2 AT MIDBAND
160	212′ 4″
80	107′ 7″
40	56′ 5″
20	28′ 5.5″
15	19′ 0″
10	14′ 0″

Fig. 2-6. Folded dipole.

additional lobes appear and change the radiation pattern (see Figs. 1-14A, 1-14B, and 1-14C, Chapter 1).

The feeders have standing waves on them because these 600-ohm feeders are connected to the center of the antenna, and the resulting impedance mismatch produces reflections. The feeders do not radiate, however, because the currents in them are equal and opposite and the resulting fields cancel each other. When some method is employed to match the antenna impedance, untuned 600-ohm feeders of any length can be used. Several methods are described in the following paragraphs.

With Delta-Matched Feeders

The impedance-transforming device shown in Fig. 2-8A is a *delta section* (a and b), so called from its resemblance to the Greek letter delta, Δ. It is formed by flaring out the ends of the nonresonant feeder wires to achieve a correspondingly gradual change in impedance. The 600-ohm feeders can be any desired length.

The delta is arranged about the center of the radiator; i.e., the center of the radiator wire is at the $a/2$ point. But the radiator does not have to be cut to insert the feeders, as it does in the two center-fed antennas described previously. The delta dimensions are:

$$a = 118/f \text{ and } b = 148/f \qquad (2\text{-}3)$$

where a and b are in feet, and f is in megahertz.

Figure 2-8B gives the a and b footages for the center frequency in each band from 10 to 40 meters (see Table 2-2, column 2 for center frequencies). Dimensions are not given for 80 and 160 meters, since the delta becomes ungainly at those wavelengths. (In the 160-meter band, for example, a = 62 feet, and b = 78 feet.)

With Gamma-Matching Section or T-Matching Section

Two other systems which, like the delta just described, match a half-wave antenna to a transmission line without splitting the radiator wire are shown in Fig. 2-9. Each of these (the *gamma-match* section in Fig. 2-9A and the *T-match* section in Fig. 2-9B) involves a conductor running parallel to the radiator for a distance a, separated from the radiator by distance b, and connected to the radiator at points at which the radiator impedance matches that of the feeders. The advantage of these matches over the delta is the elimination of the long vertical section required in the delta at some frequencies (distance b in either the gamma or T can be quite small, allowing

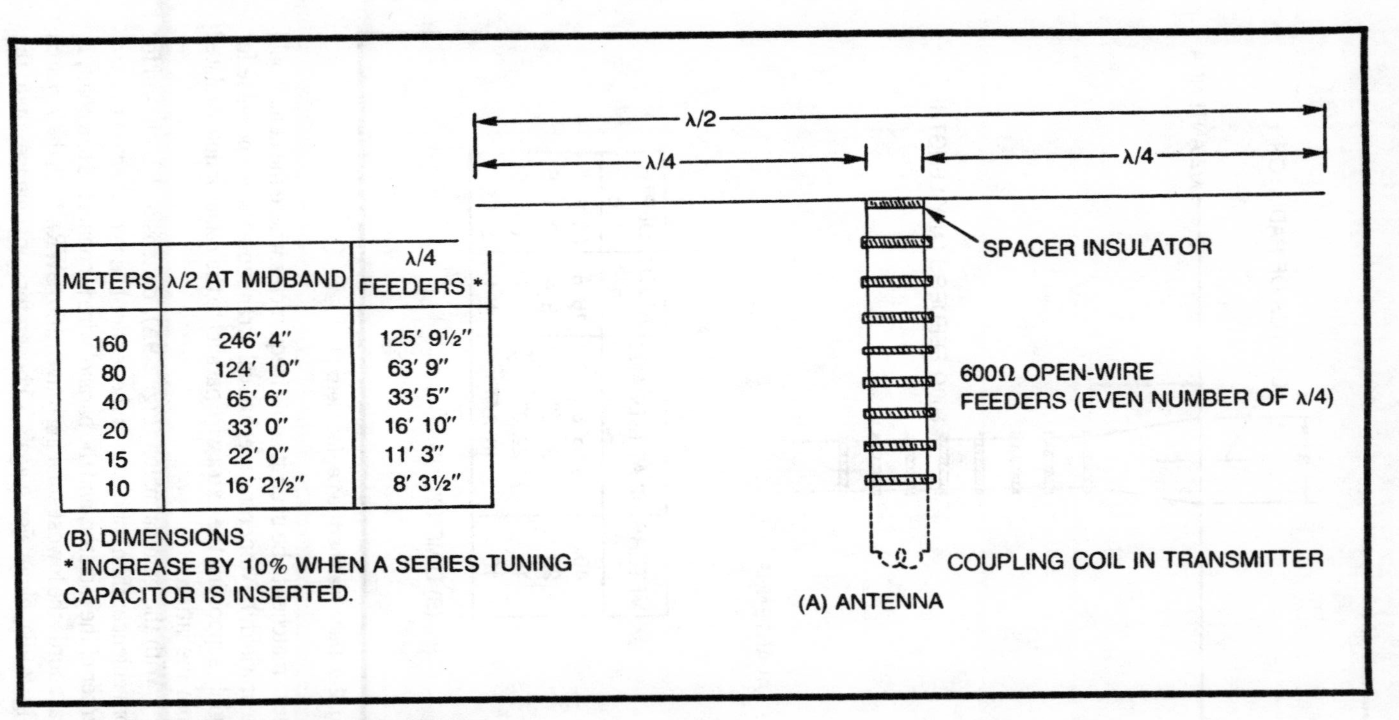

METERS	λ/2 AT MIDBAND	λ/4 FEEDERS *
160	246′ 4″	125′ 9½″
80	124′ 10″	63′ 9″
40	65′ 6″	33′ 5″
20	33′ 0″	16′ 10″
15	22′ 0″	11′ 3″
10	16′ 2½″	8′ 3½″

(B) DIMENSIONS

* INCREASE BY 10% WHEN A SERIES TUNING CAPACITOR IS INSERTED.

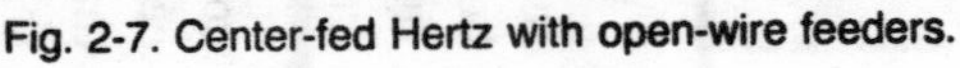

Fig. 2-7. Center-fed Hertz with open-wire feeders.

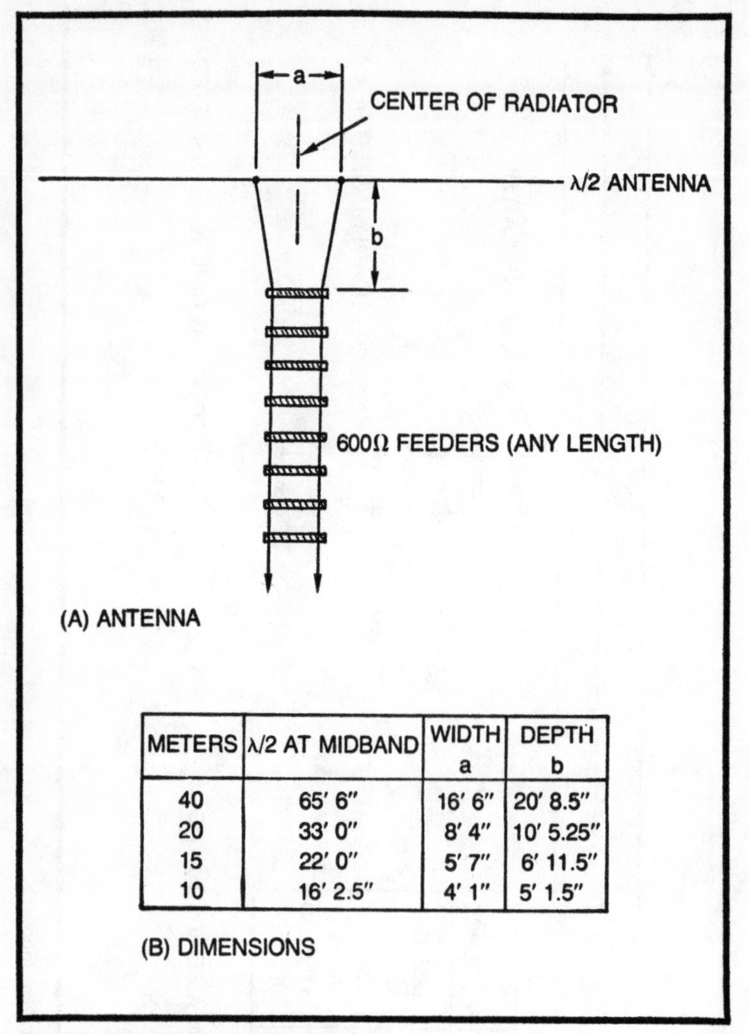

METERS	λ/2 AT MIDBAND	WIDTH a	DEPTH b
40	65′ 6″	16′ 6″	20′ 8.5″
20	33′ 0″	8′ 4″	10′ 5.25″
15	22′ 0″	5′ 7″	6′ 11.5″
10	16′ 2.5″	4′ 1″	5′ 1.5″

Fig. 2-8. Delta-matched center-fed Hertz.

these matches to be used at some frequencies where the delta would be unwieldy). Numerous mechanical arrangements are possible for rigidly supporting the extra wire parallel to the radiator and insulated from the latter.

With the gamma match (Fig. 2-9A), dimension *b* is chosen for convenience; then, one side of the feeder line is connected to exact center of the radiator, and the parallel wire is connected to a point, *1*, that affords the lowest standing-wave ratio (SWR). With the *T* match (Fig. 2-9B), dimension *b* again is chosen for convenience; then, the two parallel wires are connected to points *1* and *2*, equidistant from exact center of the radiator, that afford the lowest SWR. With either the gamma or the *T*, the length of the radiator must be pruned while the above adjustments are made, to minimize reactive components. When the radiator length and the *a* dimensions are just right at the operating frequency, the SWR will be closest to 1:1.

With Q-Bar Matching Section

In Fig. 2-10A, a linear transformer is employed to match 600-ohm nonresonant feeders to the 72-ohm impedance at the center of a

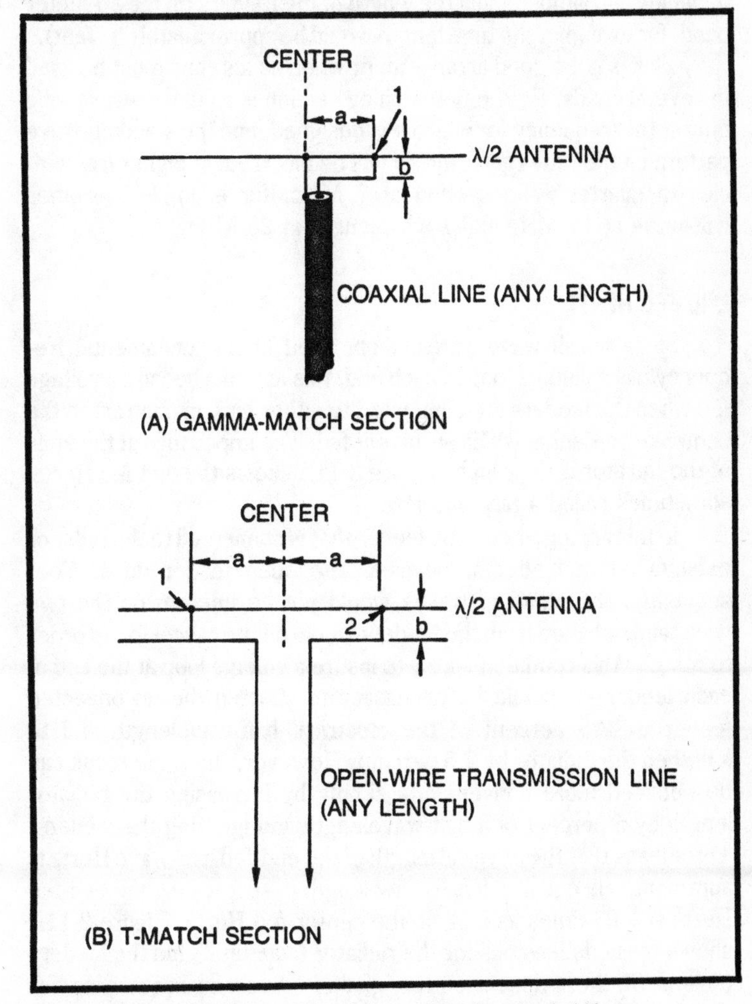

Fig. 2-9. Gamma-section and T-section matched center-fed Hertz.

half-wave antenna. This transformer consists of two parallel, quarter-wave-long, half-inch-diameter aluminum tubes (called *Q-bars*) properly spaced.

To match 600 ohms to 72 ohms, the Q-section must have a characteristic impedance of 208 ohms; for this impedance, the tubes must be spaced 1.4375 inches on centers [$208 = \sqrt{(600 \times 72)}$]. Column 3 in Fig. 2-10B shows the quarter-wavelength for the tubes at the center frequency in each band from 10 to 20 meters (see Table 2-2, column 2 for center frequencies). Dimensions are not given for the lower frequency bands since there the tube lengths become unwieldy and almost equal the length of the feeders (in the 80-meter band, for example, the tube length would be approximately 31 feet).

This is not a good arrangement when an antenna must be used in several bands, for the Q-matching section is a quarter-wave long only at the frequency for which it is designed, and the standing-wave pattern on the tubes accordingly will be incorrect at higher frequencies (a quarter-wave section at 7 MHz, for example, becomes half-wave at 14 MHz and 1 wavelength at 28 MHz).

END-FED HERTZ

Since a half-wave antenna operated at its fundamental frequency has a voltage loop at each end, this antenna becomes voltage fed when the feeders are connected to either end, in contrast to the center-fed antenna, which is current fed. The impedance at the ends of the radiator is very high. Figure 2-11A shows the end-fed Hertz, sometimes called a *zepp* antenna.

In this arrangement, the feeder that is connected to the radiator exhibits no end effects, whereas the open feeder does. Consequently, the standing waves would not be uniform on the two wires and radiation from the feeders would not be canceled. In order to remedy this condition—i.e., to insure a voltage loop at the end of each feeder—it would be necessary to shorten the unconnected feeder by 2.5 percent of the electrical half-wavelength and to lengthen the radiator by 2.5 percent. However, the same result can be obtained more conveniently simply by increasing the radiator length by 5 percent of a half-wavelength and ignoring the feeders. This means that the physical length of the end-fed half-wave Hertz is approximately equal to a half-wavelength in space; i.e., the end-fed Hertz is 1.05 times as long as the center-fed Hertz. Figure 2-11B shows these dimensions for the radiator (column 2) and the feeders (column 3); these dimensions are given for midband frequencies (for these specific frequencies, see Table 2-2, column 2). The physical

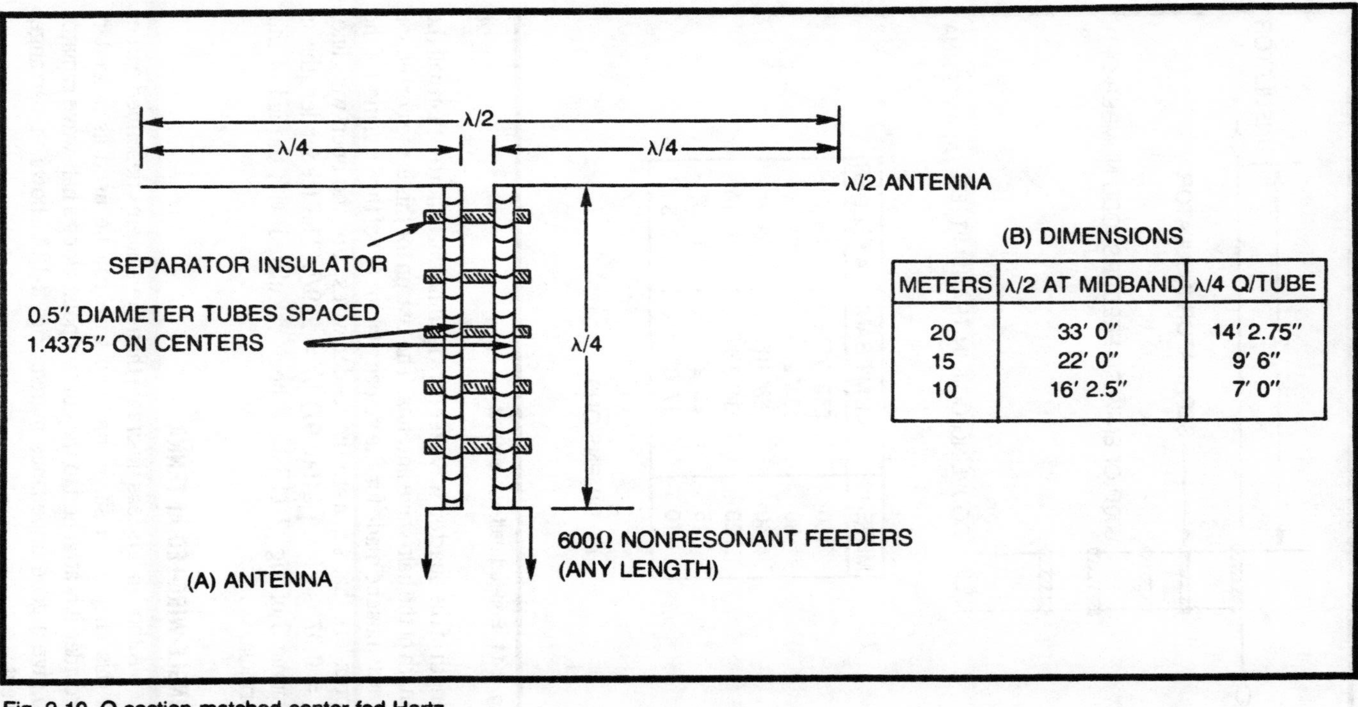

METERS	λ/2 AT MIDBAND	λ/4 Q/TUBE
20	33' 0"	14' 2.75"
15	22' 0"	9' 6"
10	16' 2.5"	7' 0"

Fig. 2-10. Q-section matched center-fed Hertz.

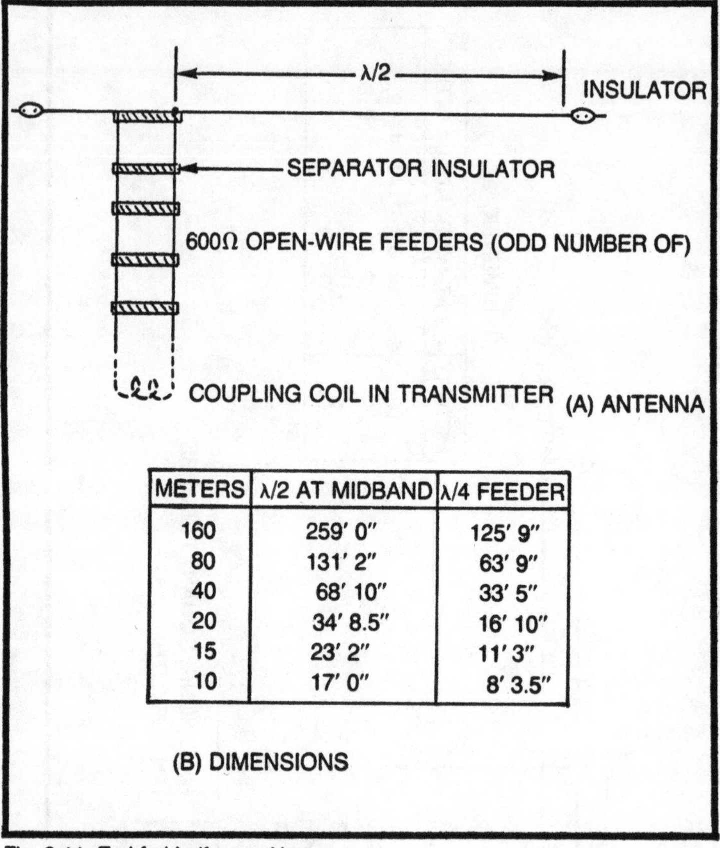

METERS	λ/2 AT MIDBAND	λ/4 FEEDER
160	259′ 0″	125′ 9″
80	131′ 2″	63′ 9″
40	68′ 10″	33′ 5″
20	34′ 8.5″	16′ 10″
15	23′ 2″	11′ 3″
10	17′ 0″	8′ 3.5″

Fig. 2-11. End-fed half-wave Hertz.

length of the quarter-wave feeders is somewhat shorter than half the length of the half-wave antenna. The reason for this is the correction which must be made for the velocity factor (V) of the insulation of the spreaders that separate the feeder wires. For the open-wire line, $V = 0.97$, so $l = [\,246(0.97)\,]/f = 239/f$. The feeder lengths in column 3 of Fig. 2-11B have been calculated with the aid of this formula.

SINGLE-WIRE-FED ANTENNA

Although it is less desirable than the systems described earlier in this chapter, a single untuned wire can be used as a feeder, provided it is attached to the correct point along a half-wave radiator to give a good impedance match. Fig. 2-12A shows this arrangement.

In practice, the single-wire feeder is connected to a point which results in the lowest SWR. This point is situated at a distance *a* approximately 15.5 percent of a half-wavelength from the center of the radiator, but must be exactly located by cut and try. Column 3 in Fig. 2-12B gives the approximate distance from the center for the center frequency in each band (for the midband frequencies, see Table 2-2, column 2).

The single-wire-fed antenna works capacitively against ground, and an actual ground connection is not mandatory. This type of antenna has two notable shortcomings: (1) the feeder does some radiating, and (2) the antenna works on its fundamental and harmonics simultaneously and accordingly demands a harmonic filter or a harmonic-attenuating coupler.

To prevent interaction between the radiator and feeder, the feeder must be run perpendicularly away from the radiator for a distance equal to at least 33 percent of a half-wavelength at the operating frequency (i.e., a distance of 81 feet 3 inches for 160

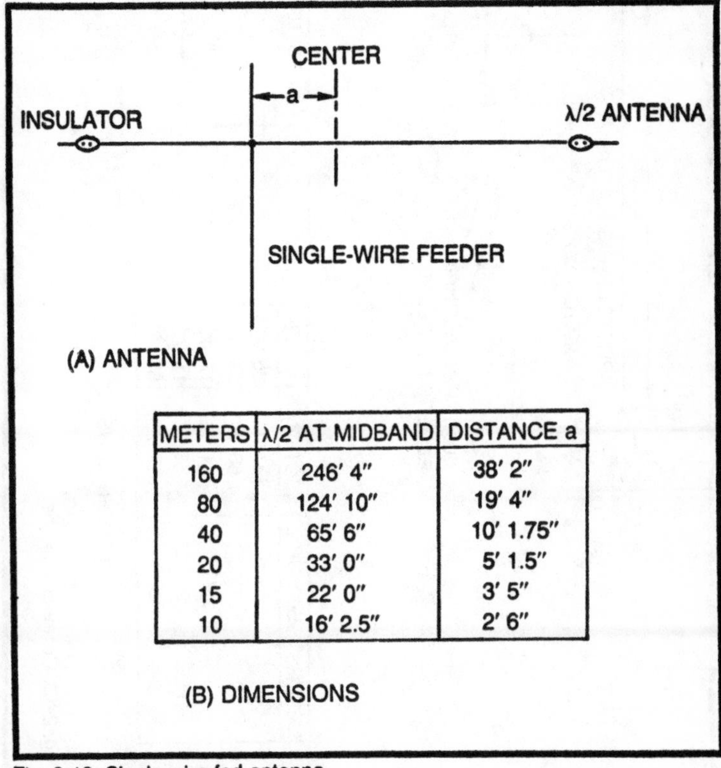

METERS	λ/2 AT MIDBAND	DISTANCE a
160	246′ 4″	38′ 2″
80	124′ 10″	19′ 4″
40	65′ 6″	10′ 1.75″
20	33′ 0″	5′ 1.5″
15	22′ 0″	3′ 5″
10	16′ 2.5″	2′ 6″

(B) DIMENSIONS

Fig. 2-12. Single-wire-fed antenna.

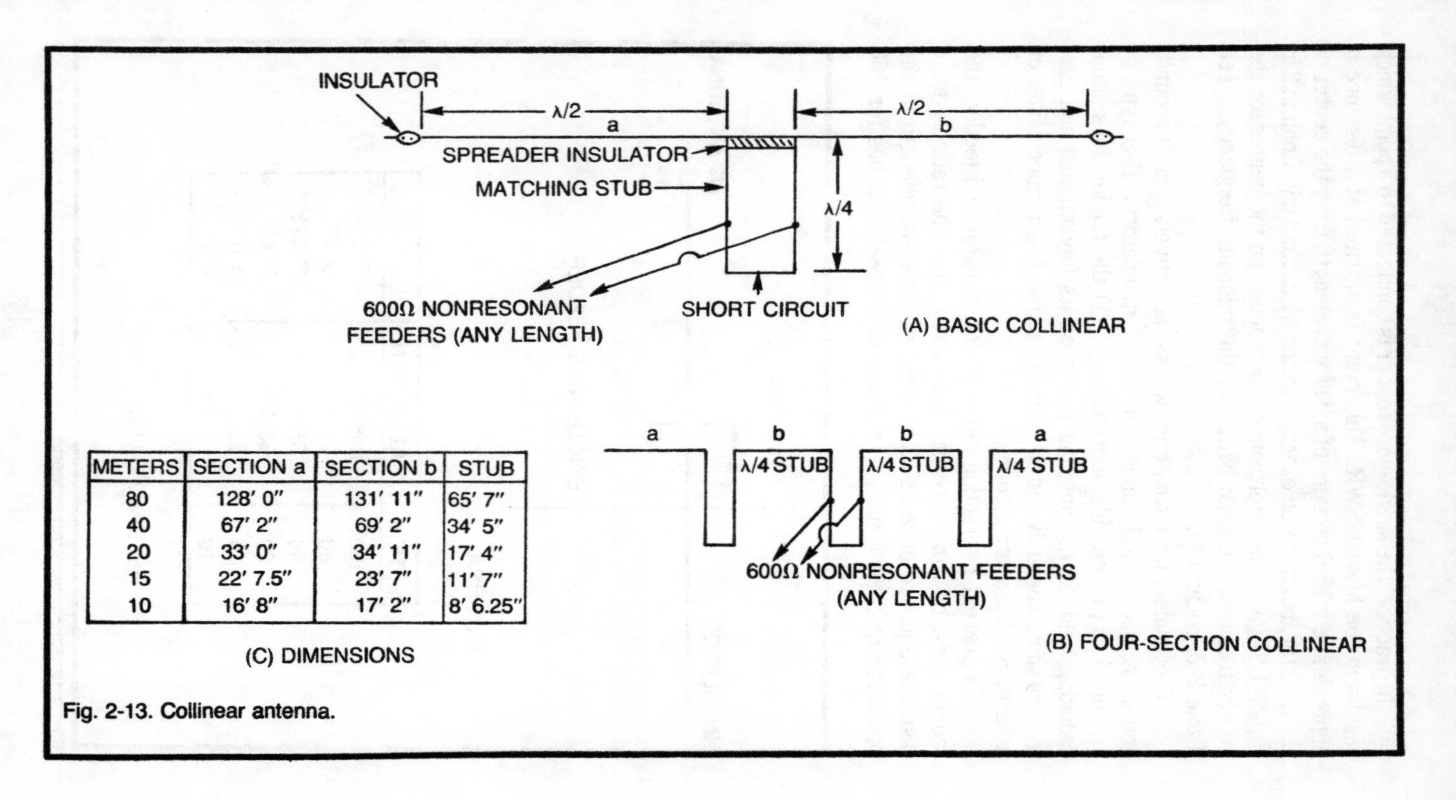

METERS	SECTION a	SECTION b	STUB
80	128′ 0″	131′ 11″	65′ 7″
40	67′ 2″	69′ 2″	34′ 5″
20	33′ 0″	34′ 11″	17′ 4″
15	22′ 7.5″	23′ 7″	11′ 7″
10	16′ 8″	17′ 2″	8′ 6.25″

(C) DIMENSIONS

Fig. 2-13. Collinear antenna.

meters, 41 feet 2 inches for 80 meters, 21 feet 7 inches for 40 meters, 10 feet 10.75 inches for 20 meters, 7 feet 3 inches for 15 meters, and 5 feet 4.25 inches for 10 meters). Also, there must be no sharp bend anywhere in the feeder.

COLLINEAR ANTENNA

The *collinear* antenna, like the various versions of halfwave dipole, radiates bidirectionally; i.e., the radiation pattern consists of two lobes (see Fig. 1-13, Chapter 1). However, the basic collinear antenna provides a gain of better than 2 dB over the dipole, and this is equivalent to multiplying the transmitter output power 1.58 times. More complicated versions of this antenna afford higher gains. Therefore, the collinear antenna is desirable wherever there is room for its erection.

The collinear antenna consists of two or more half-wave sections in phase; i.e., the current in each section is inphase with that in the other section or sections. The required phasing is taken care of by a quarter-wave stub between each two radiators. Fig. 2-13A shows the basic collinear antenna. Here, two half-wave sections in line are connected, at the beginning of each to a short-circuited quarter-wave line (AWG #12 parallel wires spaced 6 inches on centers). This quarter-wave section functions as a matching transformer (stub) with the radiator sections connected to its open ends, and a 600-ohm untuned feeder (AWG #12 wires spaced 6 inches on centers, any length) tapped along the quarter-wave section at the point resulting in maximum power transfer (lowest SWR). This arrangement is seen to consist essentially of two half-wave, end-fed antennas in line, with current flowing in the same direction in each antenna.

Figure 2-13B shows a collinear antenna consisting of four sections; this antenna provides a gain of approximately 5 dB (equivalent to multiplying the transmitter power by approximately 3.2). Each pair of sections added to the basic collinear provides additional gain.

In collinear antennas, the outside radiators (*a* in Fig. 2-13A and 2-13B) are approximately 2.6 percent longer than a physical half-wave (i.e., the length of each is $1.026 \times 468/f$), and the inside radiators (*b* in Fig. 2-13B) are approximately 5.7 percent longer than a physical half-wave (i.e., the length of each is $1.057 \times 468/f$). Similarly, the quarter-wave stubs are approximately 2.9 percent longer than a physical quarter-wave (i.e., the length of the stub is $1.029 \times 239/f$). On this basis, Fig. 2-13C gives radiator and stub dimensions for midband frequencies in five bands (for the actual frequencies, see Table 2-2, column 2). The 80-meter band is the

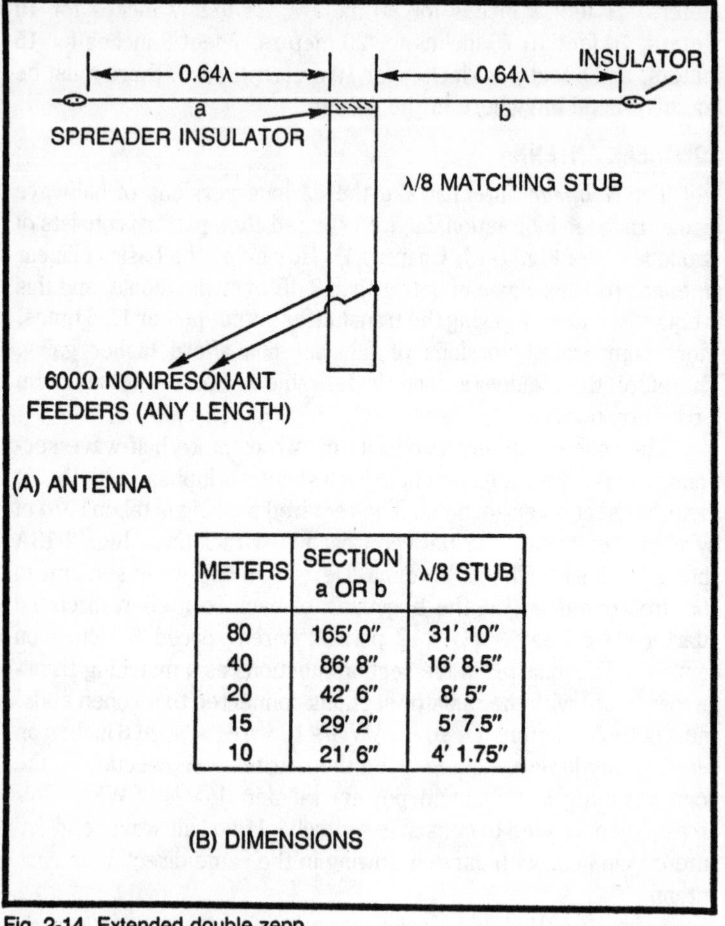

METERS	SECTION a OR b	λ/8 STUB
80	165′ 0″	31′ 10″
40	86′ 8″	16′ 8.5″
20	42′ 6″	8′ 5″
15	29′ 2″	5′ 7.5″
10	21′ 6″	4′ 1.75″

Fig. 2-14. Extended double zepp.

lowest frequency band listed here, since there is rarely enough room for a 160-meter collinear antenna (about 505 feet even for the basic, two-section antenna).

EXTENDED DOUBLE-ZEPP ANTENNA

The *extended double-zepp* antenna (Fig. 2-14A), so called because it is a length-extended version of the basic collinear antenna (Fig. 2-13A), which is in fact a double zepp, resembles the basic collinear, except that each of the sections of the extended double zepp is 0.64 wavelength long, and the matching stub is approximately an eighth-wavelength long. Fig. 2-14B gives radiator and stub dimensions. The gain of this antenna is approximately 4 dB (equivalent to multiplying the transmitter power by 2.5); thus, in this respect the extended double zepp is equivalent to a three-section collinear antenna.

As in the other collinear antennas, the matching stub consists of parallel AWG #12 wires spaced 6 inches on centers. The best length of the stub must be determined by cut and try, but the eighth-wavelength values given in column 3, Fig. 2-14B will be close to final ones. The 600-ohm nonresonant feeders are connected to the stub at the point that results in the lowest SWR.

The dimensions given in Fig. 2-14B are for midband frequencies in the 10- to 80-meter bands (for the actual frequencies, see Table 2-2, column 2).

COAXIAL ANTENNA

The *coaxial* antenna is a vertical adaptation of the simple dipole; its construction is shown in Fig. 2-15A. In the coaxial antenna, the center conductor of a coaxial cable is, in effect, extended a quarter-wavelength beyond the end of the cable, and the outer conductor of the cable is connected to a quarter-wavelength-long metal sleeve that surrounds the cable a short distance. This antenna accordingly is a half-wave, center-fed unit in which one quarter-wave section is the inner-conductor extension, and the other quarter-wave section is the metal sleeve. The cable impedance may be 50 or 53 ohms.

This arrangement provides a compact vertical antenna which gives low-angle radiation. The inner-conductor extension is a rod or whip, often telescoping for close adjustment of length, and is insulated from the sleeve through the center of which the cable passes. The sleeve may be either aluminum, copper, or brass. The coaxial antenna is usually erected by fastening the lower end of the sleeve to the top of a pole or mast by means of an insulating coupling having a hole for the cable to pass through (see Fig. 2-15B). This mounting preserves the slender profile which is characteristic of well-built vertical antennas.

The physical length of the half-wave structure limits the lowest frequency at which this type of vertical antenna can be used. The coaxial antenna thus is hardly practicable below the 20-meter band (where the radiator and sleeve each must be approximately 17 feet long, and the availability of metal tubing in the required length must be considered), and few locations perhaps can accommodate even this 34-foot length of radiator and sleeve.

GROUND-PLANE ANTENNA

The *ground-plane* antenna is a special type of quarter-wave vertical. Its distinguishing feature to the eye is the, usually four,

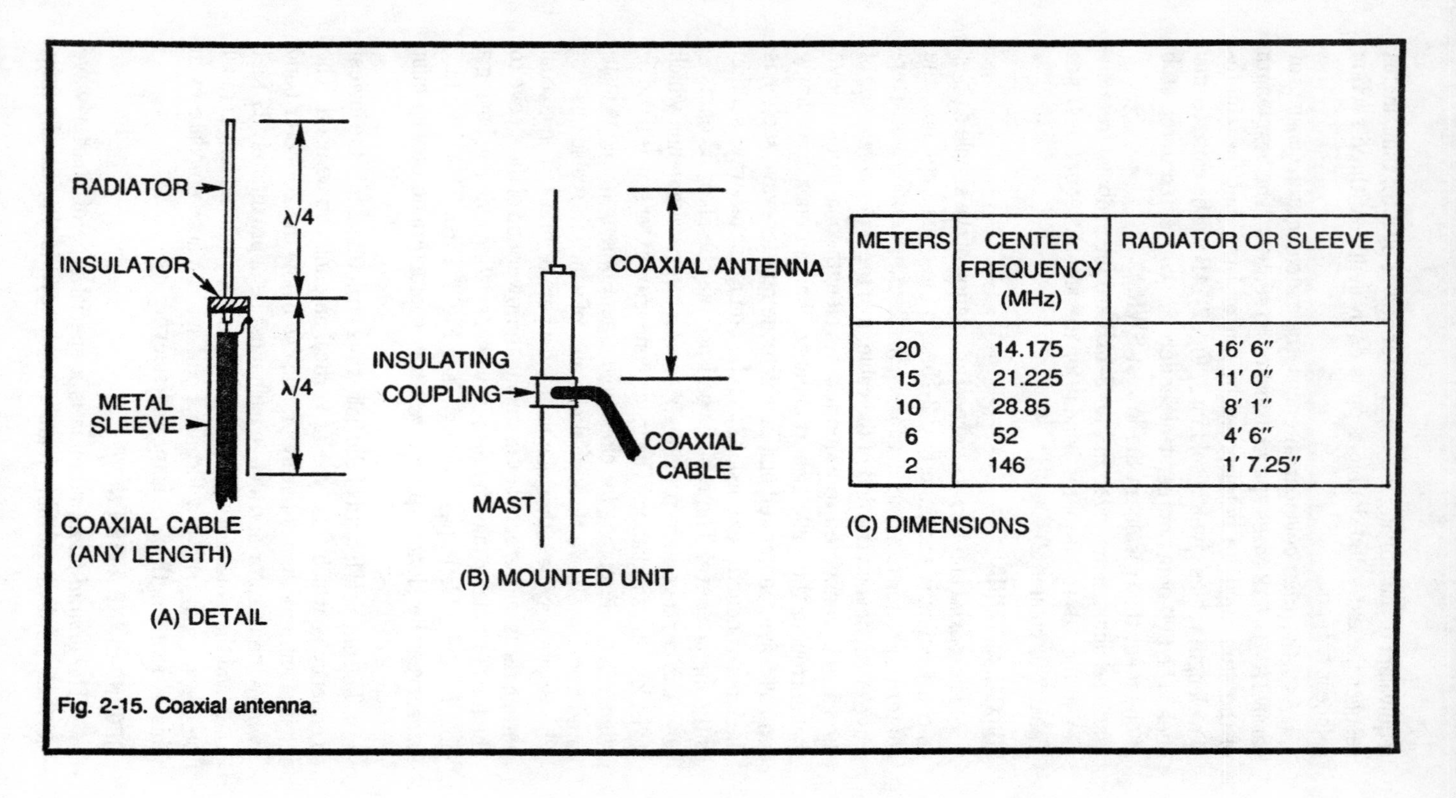

METERS	CENTER FREQUENCY (MHz)	RADIATOR OR SLEEVE
20	14.175	16′ 6″
15	21.225	11′ 0″
10	28.85	8′ 1″
6	52	4′ 6″
2	146	1′ 7.25″

(C) DIMENSIONS

Fig. 2-15. Coaxial antenna.

horizontal rods or tubes (called *radials*) that extend outward from the base of the quarter-wave radiator and are insulated from it (Fig. 2-16A). These radials, which are connected together, act somewhat as a horizontal sheet of metal would; that is, they establish a *ground plane*, an artificial, metallic ground that in effect brings ground up close to the antenna. A ground-plane antenna can be erected at any height; however, it operates best when the ground plane is at least a quarter-wavelength above the terrain. Radiation from the ground-plane antenna is omnidirectional and low angle.

The perpendicular radials are each a quarter-wavelength long and usually are self-supporting, as in Fig. 2-16A. However, slanting quarter-wave wires (or rods) insulated from the earth and the radiator are sometimes used, and these can act simultaneously as guy wires for the mast (see Fig. 2-16B). When wire radials are employed, they usually slope away from the base of the radiator at a 45° angle and are cut approximately 5 percent longer than a quarter-wavelength.

The radiator is basically a quarter-wavelength long. In some instances, however, it is composed of telescoping tubing so that the height can be adjusted closely. And while the radiator is usually tubing or rod, it sometimes is a vertical wire when support for such a wire is available.

The impedance at the base of the radiator is in the neighborhood of 30 ohms. The base can be fed from a 75-ohm coaxial line, provided a quarter-wave section of 50 ohm coax (acting as an impedance transformer) is inserted between the base and the 75-ohm line, as shown in Fig. 2-16A. Slant wires as radials, at a 45° angle (see Fig. 2-16B), raise the base impedance to approximately 50 ohms, and permit direct feed by means of 50-ohm coax, such as RG8/U. The ground-plane antenna can be matched to a given transmission-line impedance, such as 52 ohms or 75 ohms, also by varying the height of the radiator; but when this is done, the radiator will not be exactly a quarter-wavelength long in some instances. For example, the height for 52 ohms will be approximately 12 percent greater than a quarter-wavelength, and the height for 75 ohms will be approximately 24 percent greater than a quarter-wavelength. Column 3 in Fig. 2-16C gives the basic height for a quarter-wavelength, column 4 gives the height for 52 ohms, and column 5 gives the height for 75 ohms (these approximate figures are for midband frequencies in the 2- to 20-meter bands). Because the antenna is not resonant at these heights, the impedance is reactive, as well as resistive, and the reactive component must be turned out by means of a suitable capacitor or inductor at the base of the radiator.

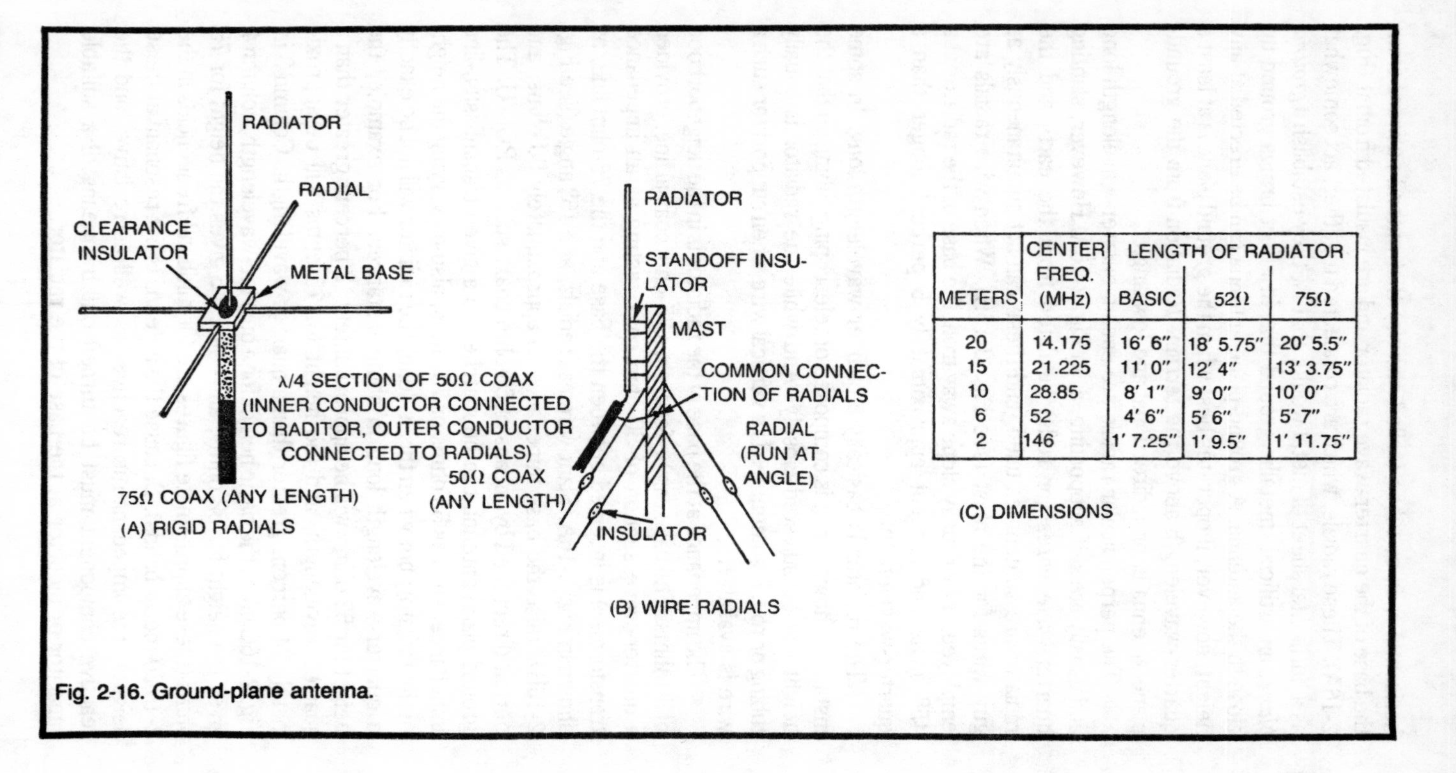

METERS	CENTER FREQ. (MHz)	LENGTH OF RADIATOR BASIC	52Ω	75Ω
20	14.175	16′ 6″	18′ 5.75″	20′ 5.5″
15	21.225	11′ 0″	12′ 4″	13′ 3.75″
10	28.85	8′ 1″	9′ 0″	10′ 0″
6	52	4′ 6″	5′ 6″	5′ 7″
2	146	1′ 7.25″	1′ 9.5″	1′ 11.75″

(C) DIMENSIONS

Fig. 2-16. Ground-plane antenna.

Because of its low angle of radiation, the ground-plane antenna is good for DX communication. Multiband operation may be achieved, provided the radials are cut to a quarter-wavelength at the lowest frequency to be used. In some installations, radiator and radials all are made telescoping for easy alteration of all lengths, as required by multiband operation.

Many types of construction are possible with the ground-plane antenna. The requirement is to insulate the radiator from the radials and to connect it to the center conductor of the coaxial line, and to connect the radials together and to the outer conductor of the line. Because of the size requirements, this antenna finds most use at 20 meters and shorter.

LONG-WIRE ANTENNA

A *long-wire* antenna is one whose radiator is longer than a half-wavelength (often, a number of wavelengths). Where there is room for its erection and when it can be supported a half-wavelength above ground, this antenna can provide some gain over a dipole. The long-wire antenna (see Fig. 2-17A) is not just a random length of wire; the length is an integral number of half-waves and is generally several wavelengths.

The long wire may be thought of as a number of half-wave sections in series. The end sections exhibit the usual end effects and their length must take these effects into consideration, but the internal sections do not show end effect, and their lengths accordingly are close to the values calculated for a half-wave in space. The result is that the total length of a given long-wire antenna is not a simple multiple of a half-wavelength, but is equal to $l = [492(N - 0.05)]/f$, where l is the length in feet, N is the number of half-waves, and f is the frequency in megahertz. For convenience, Fig. 2-17B gives lengths of long-wire antennas for 1, 2, 3, 4, and 5 wavelengths at the center frequency of all bands from 6 to 160 meters. At a given operating frequency, the gain increases with the length of the antenna.

The long-wire antenna is most effectively voltage fed at one of its ends, as shown in Fig. 2-17A. It is possible also to insert a coaxial feeder at any of the current loops along the wire; but this feeder cannot be inserted at a current node, since this would disturb the normal out-of-phase relations between currents in the adjacent half-wave sections of the wire.

Whereas maximum radiation occurs broadside to a dipole, it occurs off the ends of a long wire. If the far end is terminated with a

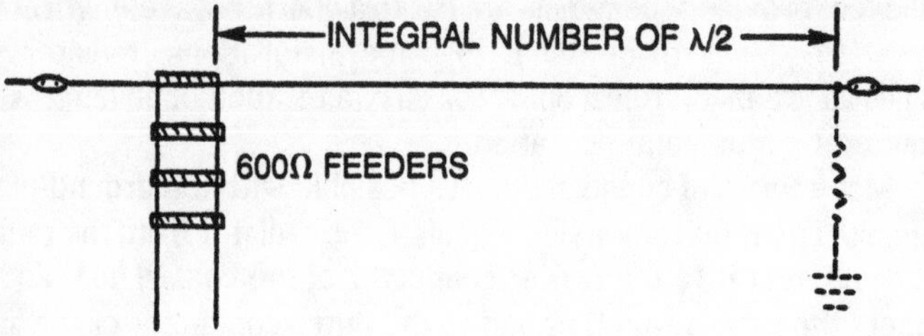

INTEGRAL NUMBER OF $\lambda/2$

600Ω FEEDERS

(A) DETAIL

METERS	CENTER FREQ. (MHz)	PHYSICAL LENGTH OF RADIATOR				
		λ	2λ	3λ	4λ	5λ
160	1.9	505′ 0″	1023′ 0″	1541′ 0″	2059′	2576′
80	3.75	255′ 0″	518′ 0″	781′ 0″	1043′	1305′
40	7.15	134′ 0″	272′ 0″	409′ 0″	547′ 0″	685′ 0″
20	14.175	67′ 8″	137′ 0″	206′ 0″	276′ 0″	345′ 0″
15	21.225	45′ 0″	91′ 7″	138′ 0″	184′ 0″	231′ 0″
10	28.85	33′ 3″	67′ 4″	101′ 6″	135′ 7″	169′ 8″
6	52	18′ 5″	37′ 4″	56′ 3″	75′ 3″	94′ 2″

(B) DIMENSIONS

Fig. 2-17. Long-wire antenna.

noninductive resistance equal to the characteristic impedance of the antenna (approximately 90 ohms at 1 wavelength to 135 ohms at 5 wavelengths), maximum radiation is from the far end only. Thus, the unterminated long-wire antenna is end-fire bidirectional, and the terminated long-wire is end-fire unidirectional. Although there is a great deal of power in the main lobes (approximately 4 dB gain at 5 wavelengths), the long wire radiates many minor lobes that increase in number with the length of the wire. It is interesting to note that the gain provided by the long-wire antenna at 10 wavelengths is equivalent to multiplying the transmitter power by 5.5.

All harmonically operated antennas become long-wire antennas at some frequency. Thus, when operated at 10 meters, an 80-meter half-wave antenna becomes a 4-wavelength long wire. At very long lengths (15 wavelengths and longer), the response of a long wire becomes broadband, requiring little, if any, tuning.

V-ANTENNA

When two identical long-wire antennas are mounted at an angle with respect to each other and end fed simultaneously as if they were the two halves of a center-fed Hertz, the result is a *V-antenna* (see Fig. 2-18A), so called from its resemblance to the letter *V*. In this antenna, the main lobes of the long-wire sections reinforce each other, and maximum radiation and reception occur in the forward and backward direction along the line bisecting the angle of the *V*, as shown by the arrow in Fig. 2-18A. This antenna thus is bidirectional.

Figure 2-18C gives the required physical length (l) of each leg of the *V*, and the corresponding angle (A) for 1, 2, 3, 4, and 5 wavelengths at the center frequencies in the 6- to 40-meter bands. The angle a in each instance is equal to twice the angle of radiation of a long-wire antenna of length l; this permits maximum combining of the lobes of the separate wires of the *V*. Since each section of the antennas listed in Fig. 2-18C is an even number of quarter-waves long, voltage feed must be employed (current feed is permissible where l is an odd number of quarter-waves).

The gain of the V-antenna varies directly with length l, and inversely with angle a; it reaches approximately 11 dB, for example, when $a = 30°$. The longer the V-antenna, the higher its gain and the sharper its directivity, and vice versa. The best height for this type of horizontal antenna is a half-wavelength to 1 wavelength above ground.

RHOMBIC ANTENNA

The *rhombic* antenna consists, in effect, of four long wires connected in the shape of a rhombus or diamond. It resembles two V-antennas connected open end to open end. This horizontal antenna is unidirectional and also is nonresonant, in both transmitting and receiving.

The basic arrangement of the rhombic antenna is shown in Fig. 2-19A. The four legs of the antenna have equal length, l; and the side angle (a) between the two legs in one section is equal to the side angle (a) between the two legs in the other section. This antenna exhibits unidirectivity, nonresonance, and maximum gain when its free end is terminated with a 750-ohm noninductive resistor, R, capable of handling the transmitter power. The nonresonant feeders

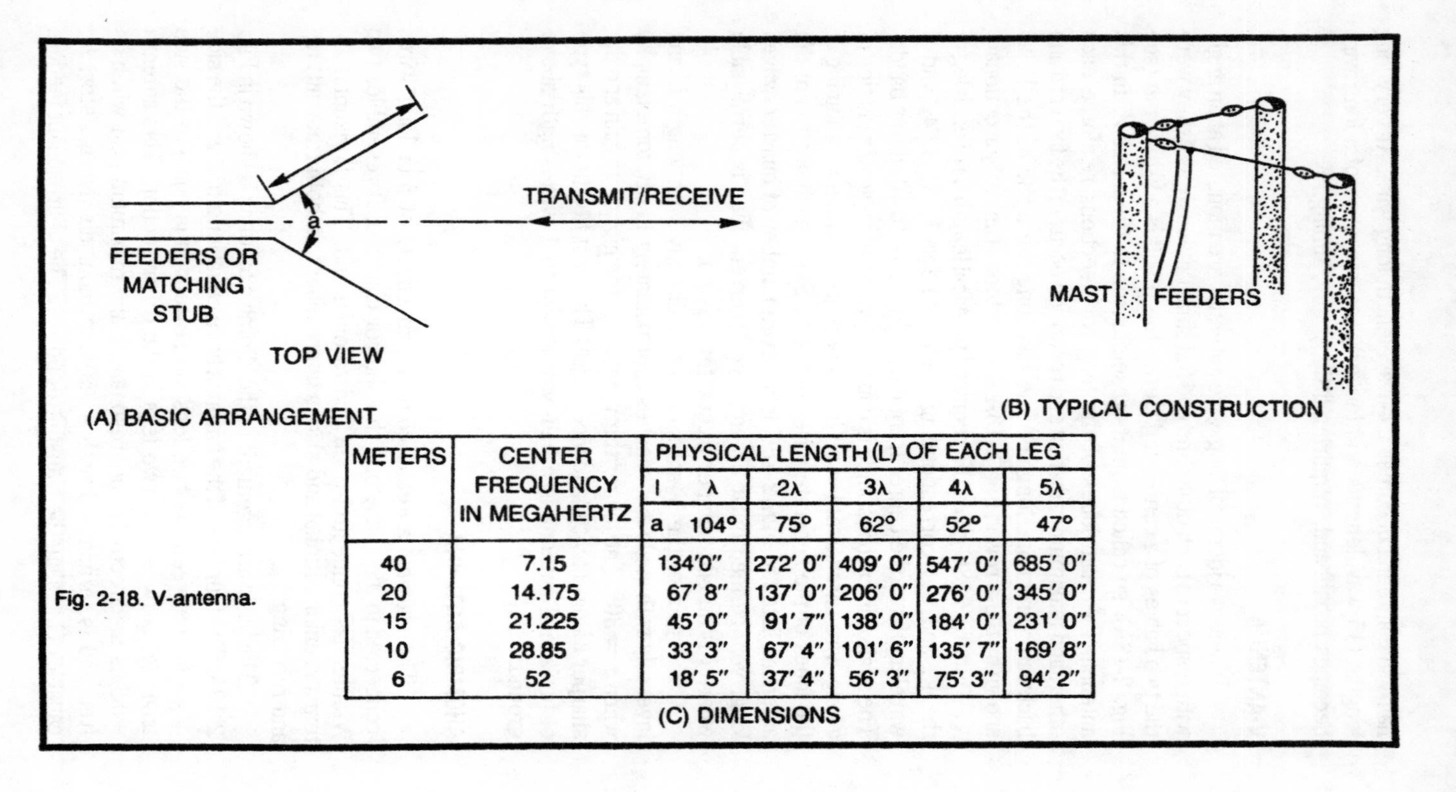

METERS	CENTER FREQUENCY IN MEGAHERTZ	PHYSICAL LENGTH (L) OF EACH LEG				
		l λ	2λ	3λ	4λ	5λ
		a 104°	75°	62°	52°	47°
40	7.15	134'0"	272' 0"	409' 0"	547' 0"	685' 0"
20	14.175	67' 8"	137' 0"	206' 0"	276' 0"	345' 0"
15	21.225	45' 0"	91' 7"	138' 0"	184' 0"	231' 0"
10	28.85	33' 3"	67' 4"	101' 6"	135' 7"	169' 8"
6	52	18' 5"	37' 4"	56' 3"	75' 3"	94' 2"

(C) DIMENSIONS

Fig. 2-18. V-antenna.

should be approximately 700 ohms (for example, two AWG #12 wires spaced 13.75 inches on centers).

Figure 2-19C shows required physical length l of each leg of the rhombic, and the corresponding side angle a for one, two, three, four, and five wavelengths at the center frequency in the 6- to 20-meter bands. In the rhombic antenna, the lengths are not critical, so they are given only to the nearest foot in Fig. 2-19C.

A rhombic antenna should be erected at least a half-wavelength above ground at the lowest operating frequency, and should be kept horizontal (see Fig. 2-19B). Directivity suffers when the antenna is tilted. Figure 2-19B depicts the usual type of installation.

Where there is room for its erection, the rhombic antenna is often the most effective nonrotating directive (beam) antenna for ham use. A gain of over 10 dB is common.

TWO-ELEMENT BEAMS

An undriven dipole mounted parallel to and 0.10 to 0.25 wavelength behind a regular driven dipole (the *radiator*) will be excited parasitically by the latter and will act as a reflector, intensifying the radiation in essentially one, broadside direction. This arrangement, shown in Fig. 2-20A, thus constitutes a two-element beam antenna. It belongs to the family of *Yagi* antennas, distinguished by their use of parasitic, rather than directly driven secondary elements.

As a directive antenna, the two-element unit is simple and provides a good front-to-back ratio. At 20 meters and lower, its size is such that it can be rotated, either manually or by motor, for choice of direction. Figure 2-20B depicts one type of construction for the stationary beam employing separate masts and wire elements (this arrangement would be used only on 160, 80, and 40 meters; on 20, 15, 10, and 6 meters, small compact rod or tubing assemblies are used—see Fig. 2-20C). If the radiator, reflector, and spacing are correctly dimensioned, the impedance of the dipole remains undisturbed and the radiator may be center fed with a coaxial line. With narrow spacing (down to 0.1 wavelength), however, the dipole impedance is lowered, and steps must be taken to match the coax (or other line) to the radiator.

In the two-element beam, the radiator is a standard dipole; however, the reflector is 5 percent longer than the radiator. The formulas for the two-element beam are:

$$\text{Reflector } a = 492/f \qquad (2\text{-}4)$$

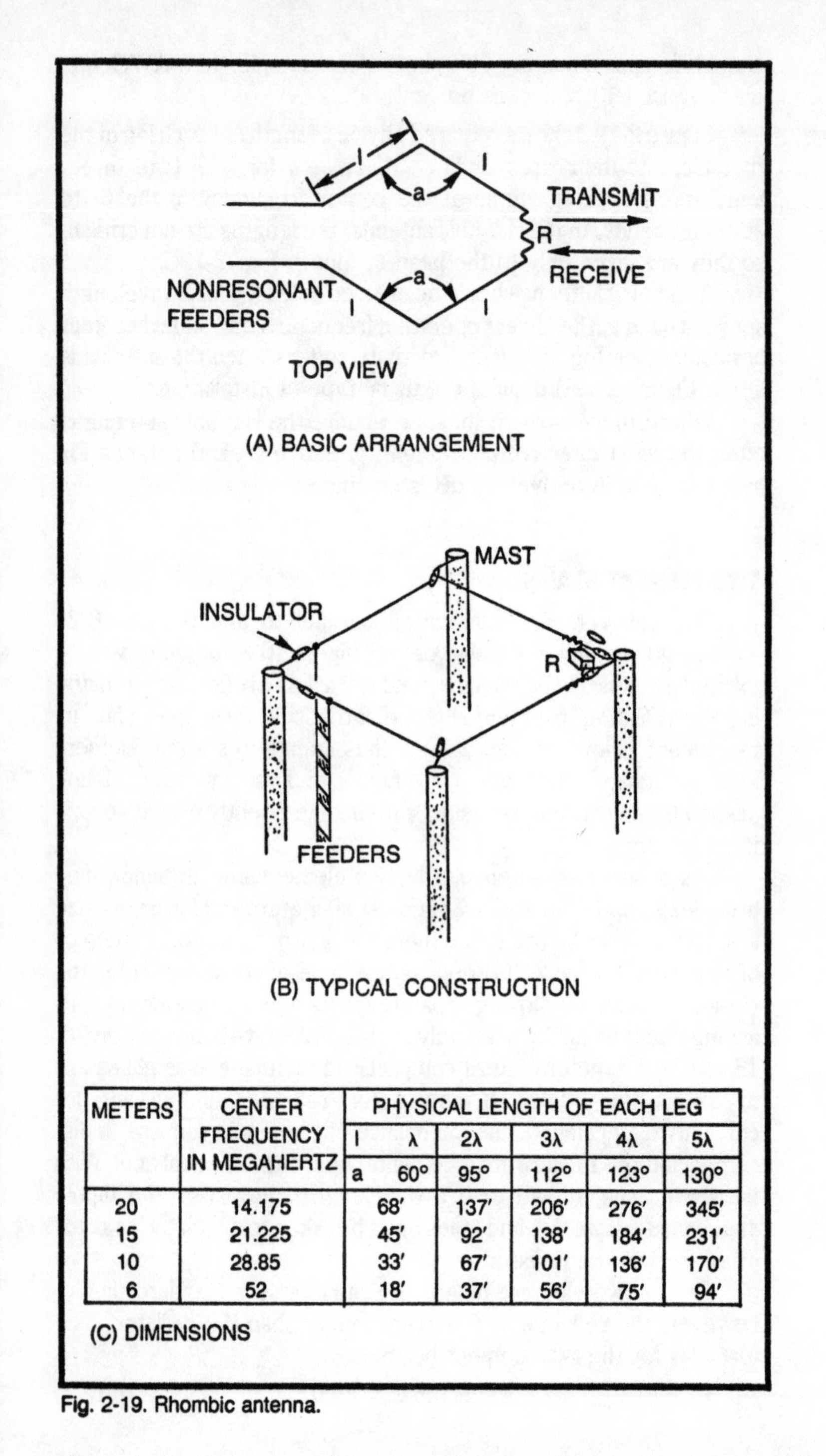

METERS	CENTER FREQUENCY IN MEGAHERTZ	PHYSICAL LENGTH OF EACH LEG				
		l λ	2λ	3λ	4λ	5λ
		a 60°	95°	112°	123°	130°
20	14.175	68′	137′	206′	276′	345′
15	21.225	45′	92′	138′	184′	231′
10	28.85	33′	67′	101′	136′	170′
6	52	18′	37′	56′	75′	94′

(C) DIMENSIONS

Fig. 2-19. Rhombic antenna.

$$\text{Spacing } b = 246/f \tag{2-5}$$

$$\text{Radiator halves } c, d = 234/f \tag{2-6}$$

All dimensions are in feet, and frequency f is in megahertz.

Figure 2-20D, for convenience, gives radiator and reflector lengths and quarter-wave spacing for center frequencies in each band from 6 to 160 meters. While wire will be used in the longer antennas (40 to 160 meters), rigid tubing or rod may be used in the shorter ones. The tubing may be made telescoping for close adjustment of lengths.

In another version of the two-element beam, the parasitic element is mounted in front of the radiator, instead of behind it. The parasitic element then becomes a ***director***. Maximum radiation still is in the same broadside direction, i.e., in a line extending from the

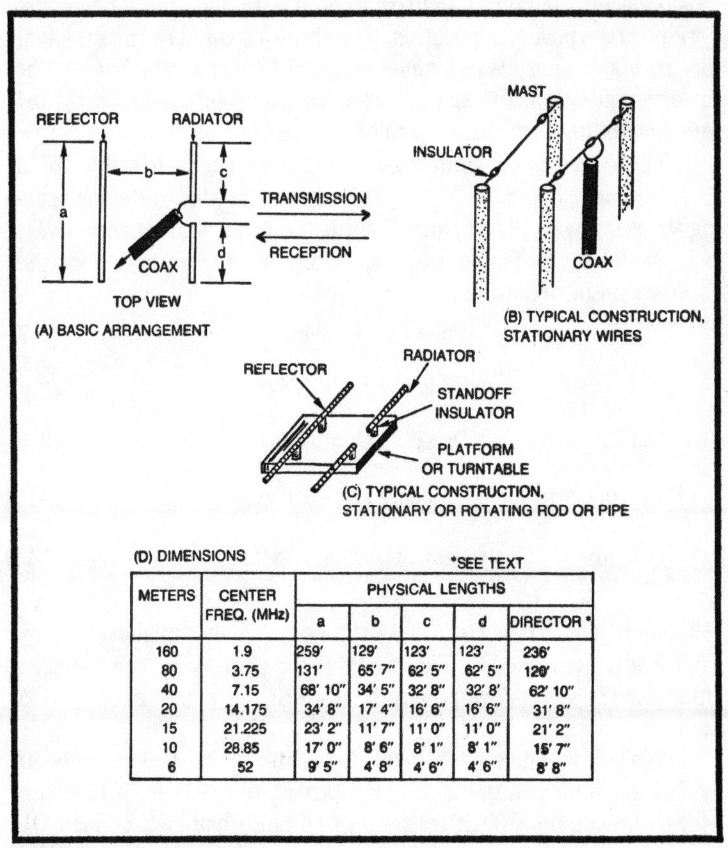

(D) DIMENSIONS

*SEE TEXT

METERS	CENTER FREQ. (MHz)	PHYSICAL LENGTHS				
		a	b	c	d	DIRECTOR *
160	1.9	259′	129′	123′	123′	236′
80	3.75	131′	65′ 7″	62′ 5″	62′ 5″	120′
40	7.15	68′ 10″	34′ 5″	32′ 8″	32′ 8′	62′ 10″
20	14.175	34′ 8″	17′ 4″	16′ 6″	16′ 6″	31′ 8″
15	21.225	23′ 2″	11′ 7″	11′ 0″	11′ 0″	21′ 2″
10	28.85	17′ 0″	8′ 6″	8′ 1″	8′ 1″	15′ 7″
6	52	9′ 5″	4′ 8″	4′ 6″	4′ 6″	8′ 8″

Fig. 2-20. Two-element beam.

radiator and through the director. A director is 4 percent shorter than the radiator (see last column in Fig. 2-20D).

At the lower ham frequencies, both the length and separation of the elements are sizable; nevertheless, the 40-, 80-, and 160-meter dimensions are included in Fig. 2-20D for the convenience of those readers who have the room for such an antenna. At the higher frequencies, where elements and separation are shorter, the antenna is often made vertical, instead of horizontal, for vertical polarization.

THREE-ELEMENT BEAM

The three-element beam shown in Fig. 2-21A is a Yagi unit that has a parasitic reflector behind the dipole radiator, and a parasitic director in front of the radiator. This antenna provides higher gain (up to 7 dB) and better front-to-back ratio (20 to 30 dB typical for close-spaced elements) than that of the two-element beam described in the preceding section. The three-element beam is popular as a rotatable antenna at frequencies of 14 MHz and higher. At the lower frequencies, the space it requires is prohibitive (259 by 181 feet for 160 meters, for example).

There are many combinations of dimensions possible for a three-element beam. In Fig. 2-21A, the radiator-to-reflector spacing (D) is 0.2 wavelength, and the radiator-to-director spacing (E) is 0.15 wavelength. These are near optimum. The formulas for the three-element beam are:

$$\text{Reflector } A = 492/f \tag{2-7}$$

$$\text{Radiator } B = 468/f \tag{2-8}$$

$$\text{Director } C = 450/f \tag{2-9}$$

$$\text{Radiator-to-reflector spacing } D = 197/f \tag{2-10}$$

$$\text{Radiator-to-director spacing } E = 148/f \tag{2-11}$$

All dimensions are in feet, and frequency f is in megahertz.

For convenience, in Fig. 2-21C these dimensions are precalculated for the center frequency of each of the bands from 6 to 20 meters.

While it is entirely feasible to construct a three-element beam with wire (a mounting boom being provided for the rotatable version), the common form consists of all pipe (half-inch or inch OD tubing) without insulation—the plumber's delight (see Fig. 2-21B).

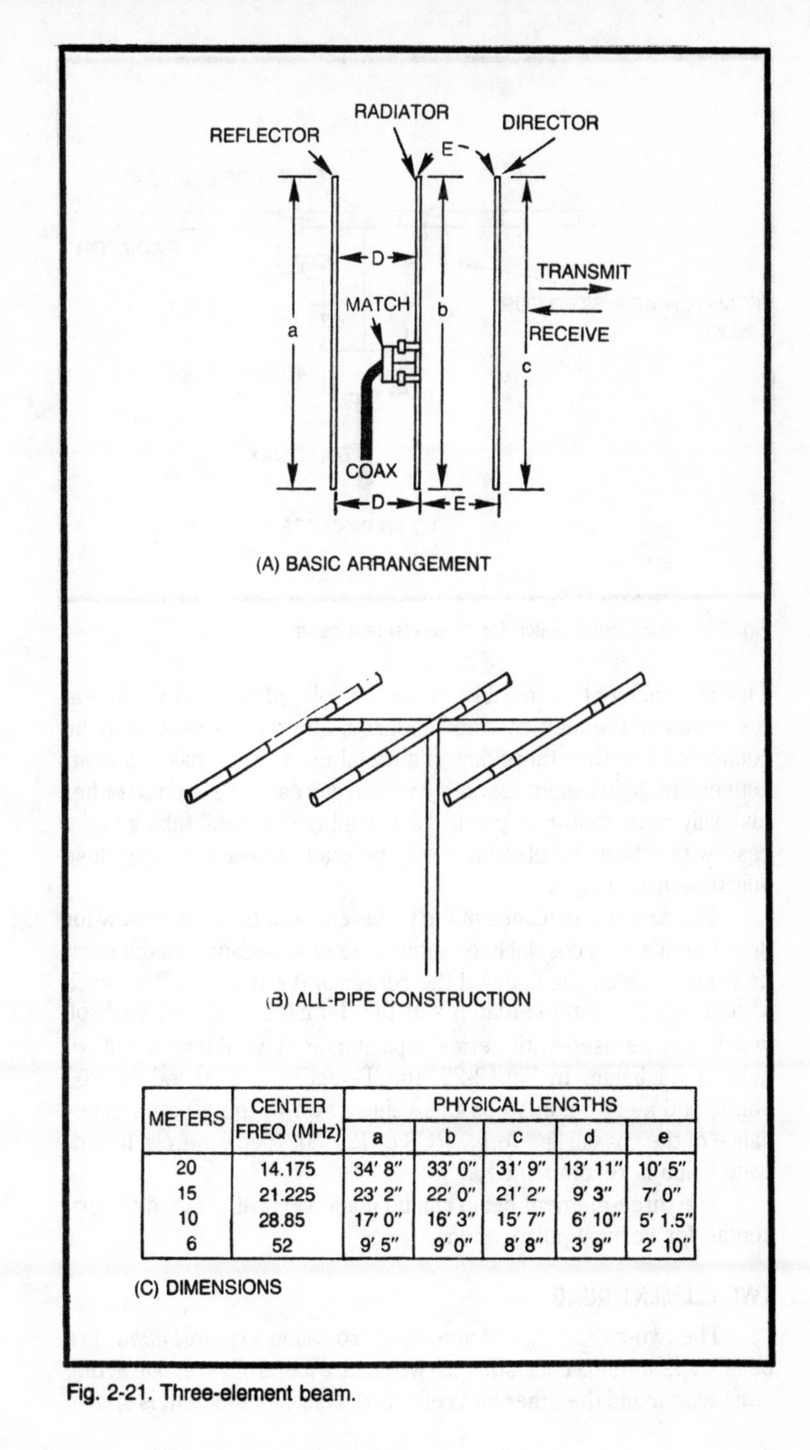

METERS	CENTER FREQ (MHz)	PHYSICAL LENGTHS				
		a	b	c	d	e
20	14.175	34′ 8″	33′ 0″	31′ 9″	13′ 11″	10′ 5″
15	21.225	23′ 2″	22′ 0″	21′ 2″	9′ 3″	7′ 0″
10	28.85	17′ 0″	16′ 3″	15′ 7″	6′ 10″	5′ 1.5″
6	52	9′ 5″	9′ 0″	8′ 8″	3′ 9″	2′ 10″

Fig. 2-21. Three-element beam.

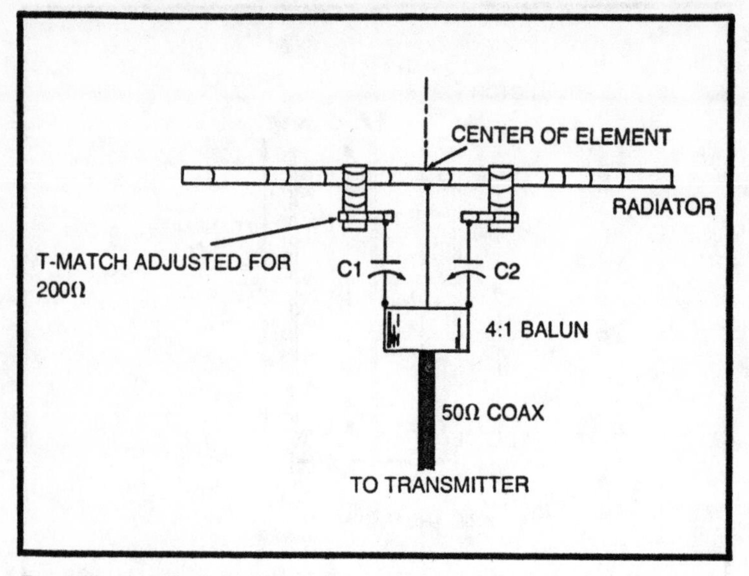

Fig. 2-22. Impedance match for three-element beam.

This arrangement is possible, since there is, ideally, no voltage at the center of the elements and consequently the elements may be connected together there and to a metal mast. This makes a neat, convenient, lightweight assembly which may easily be rotated either manually or by motor. A particular advantage of metal tubing is the ease with which the elements may be made telescoping for close adjustment of length.

The center impedance of the three-element beam is too low for direct feed with a coaxial line; some type of impedance match must be used between the line and the center of the radiator. Two such devices are the gamma-match and the T-match sections, each of which may be used with series capacitance. The T-match will require a 4:1 balun. In Fig. 2-22, the T-section is adjusted for 200 ohms, and a 4:1 balun steps this value down to the 50-ohm impedance of the coaxial line (e.g., RG8 or RG58). A commercial broadband balun is recommended.

The three-element beam can be made vertical, instead of horizontal, for vertical polarization.

TWO-ELEMENT QUAD

The two-element *quad* antenna (also called a *cubical quad*) is a beam-type antenna consisting of two square loops of wire, one acting as a radiator and the other as a reflector. This arrangement is shown in Fig. 2-23A. The radiator loop consists of a 1-wavelength-long strand of AWG #12 wire, whereas the reflector loop contains 1.05 wavelengths of wire. Separation between the loops is an eighth-wavelength. The radiator thus measures 0.25 wavelength per side, and the reflector 0.2625 wavelength per side. The formulas for the two-element quad are:

$$\text{Reflector} = 258.3/f \text{ feet per side} \tag{2-12}$$

$$\text{Radiator} = 246/f \text{ feet per side} \tag{2-13}$$

$$\text{Spacing} = 123/f \text{ feet} \tag{2-14}$$

Frequency f is in megahertz.

For convenience, Fig. 2-23B gives quad dimensions for the center frequency in the 10- to 40-meter bands. At longer wavelengths, the quad structure becomes ungainly.

The radiator is fed at the center of its bottom side with 75-ohm coaxial cable, such as RG59. The cable should be brought straight away and down from the radiator for as long a distance as practicable.

The quad wires are supported on a frame of insulating material, such as wood, bamboo, or plastic. This frame may be rotated horizontally, either manually or by motor.

THREE-ELEMENT QUAD

The three-element quad is a beam antenna consisting of three square loops of wire, one acting as a radiator, one as a reflector, and one as a director. This arrangement is shown in Fig. 2-24A. The three-element quad provides higher gain and better directivity than that obtained with a two-element quad.

In the three-element quad, the radiator loop consists of a 1-wavelength-long strand of AWG #12 wire, the reflector loop contains 1.05 wavelengths of AWG #12 wire, and the director loop contains 0.95 wavelength of wire. The reflector is spaced an eighth-wavelength behind the radiator, and the director an eighth-wavelength in front of the radiator. The radiator thus is 0.25 wavelength per side, the reflector 0.2625 wavelength per side, and the director 0.2375 wavelength per side. The formulas for the three-element quad are:

$$\text{Reflector} = 258.3/f \text{ feet per side} \tag{2-15}$$

$$\text{Radiator} = 246/f \text{ feet per side} \tag{2-16}$$

$$\text{Director} = 234/f \text{ feet per side} \tag{2-17}$$

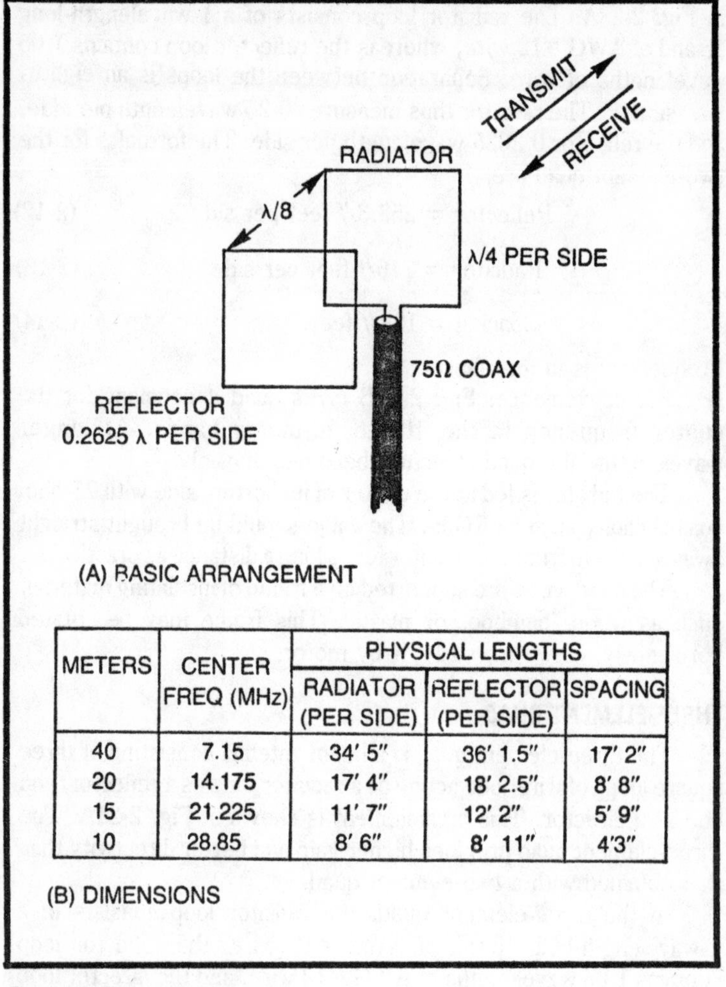

METERS	CENTER FREQ (MHz)	PHYSICAL LENGTHS		
		RADIATOR (PER SIDE)	REFLECTOR (PER SIDE)	SPACING
40	7.15	34′ 5″	36′ 1.5″	17′ 2″
20	14.175	17′ 4″	18′ 2.5″	8′ 8″
15	21.225	11′ 7″	12′ 2″	5′ 9″
10	28.85	8′ 6″	8′ 11″	4′3″

(B) DIMENSIONS

Fig. 2-23. Two-element quad.

$$\text{Spacing} = 123/f \text{ feet} \qquad (2\text{-}18)$$

Frequency f is in megahertz.

For convenience, Fig. 2-24B gives quad dimensions for the center frequency in the 10- to 40-meter bands. At longer wavelengths, the quad structure becomes unwieldy.

The radiator is fed at the center of its bottom side with 75-ohm coaxial cable, such as RG59. The cable should be brought straight away and down from the radiator for as long a distance as practicable.

Like the two-element quad, the three-element unit is supported on a frame of insulating material, such as wood, bamboo, or plastic, and this frame may be rotated horizontally, either manually or by motor.

Figure 2-25 gives data for a three-band quad antenna. This arrangement is popular for service on 10, 15, and 20 meters, as shown.

This antenna system consists of separate quad radiator units interlaced with each other, and separate quad director units interlaced with each other and placed in front of the radiators at a distance equal to an eighth-wavelength at 20 meters (see Fig. 2-25A). The radiator section is fed with 300-ohm TV ribbon connected at the bottom center of the 10-meter element and transposed between the

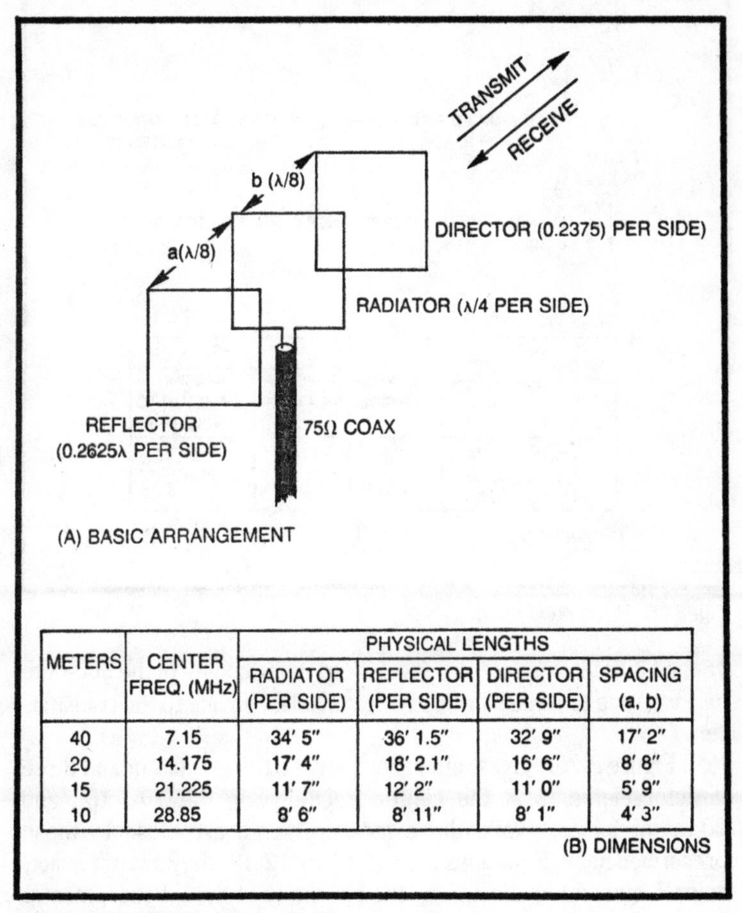

METERS	CENTER FREQ. (MHz)	PHYSICAL LENGTHS			
		RADIATOR (PER SIDE)	REFLECTOR (PER SIDE)	DIRECTOR (PER SIDE)	SPACING (a, b)
40	7.15	34′ 5″	36′ 1.5″	32′ 9″	17′ 2″
20	14.175	17′ 4″	18′ 2.1″	16′ 6″	8′ 8″
15	21.225	11′ 7″	12′ 2″	11′ 0″	5′ 9″
10	28.85	8′ 6″	8′ 11″	8′ 1″	4′ 3″

(B) DIMENSIONS

Fig. 2-24. Three-element quad.

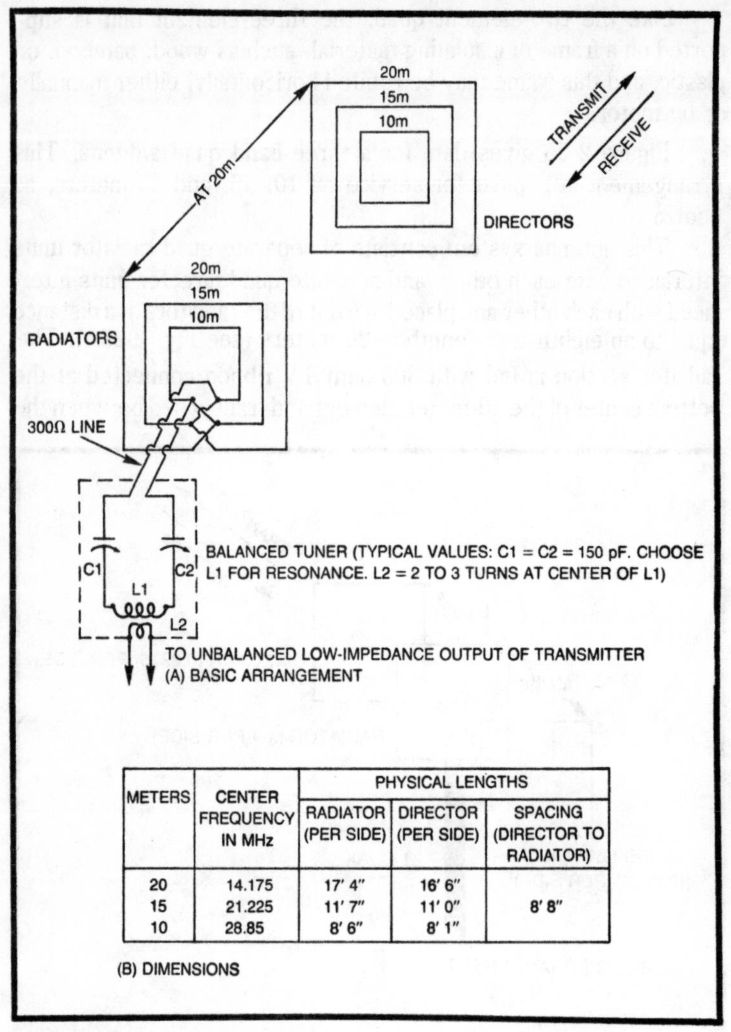

METERS	CENTER FREQUENCY IN MHz	PHYSICAL LENGTHS		
		RADIATOR (PER SIDE)	DIRECTOR (PER SIDE)	SPACING (DIRECTOR TO RADIATOR)
20	14.175	17′ 4″	16′ 6″	
15	21.225	11′ 7″	11′ 0″	8′ 8″
10	28.85	8′ 6″	8′ 1″	

(B) DIMENSIONS

Fig. 2-25. Three-band quad.

15- and 20-meter elements. A balanced tuner converts the 300-ohm balanced line to coaxial or open-wire unbalanced line to the transmitter.

Figure 2-25B gives dimensions for the three radiator and three director elements at the center frequency in the 10-, 15-, and 20-meter bands. While these measurements are close to those prescribed here, Equations 2-16, 2-17, and 2-18, they are not exact, as each element is affected by the proximity of the other elements. Hence, final pruning of length is required for best SWR.

As in the other quads described in the preceding sections, the three-band unit is supported on a frame of insulating material, such as wood, bamboo, or plastic. In this three-band quad, the frame is designed for the longest wavelength elements (here, 20 meters).

ADDITIONAL NOTES ON QUADS

The following additional notes will be useful to prospective builders and users of quad antennas.

1. Because of its low angle of radiation, the quad antenna is favored for DX transmission and reception.
2. Quads are broad tuning.
3. Generally, arrangements for holding and rotating a quad antenna are easier to build and operate than those for Yagi beams.
4. There seems to be little difference in results between arranging the elements square—as shown in Figs. 2-23, 2-24, and 2-25—or diamond (standing on one corner).
5. Additional director elements may be used for improved performance.
6. If, for mechanical reasons, it is desired to keep the radiator and reflector of the same physical size, the reflector then must be tuned with a shorted stub or a loading coil, either of which must be inserted into the reflector at the center of its bottom side.
7. In the three-band unit, each radiator and its associated director should be adjusted separately for maximum gain in the forward direction (see arrow in Figs. 2-23A, 2-24A, and 2-25A).
8. The forward gain and the front-to-back ratio of a quad do not necessarily occur concurrently. Some compromise therefore is necessary when this type of antenna is adjusted for optimum performance.

LIMITED-SPACE ANTENNAS

It is very common for hams living in densely populated areas to be cramped for antenna space. This situation has been eased in a number of ingenious ways; though the methods employed are too numerous to be described in total here, several limited-space antennas of proven worth are presented in the following paragraphs.

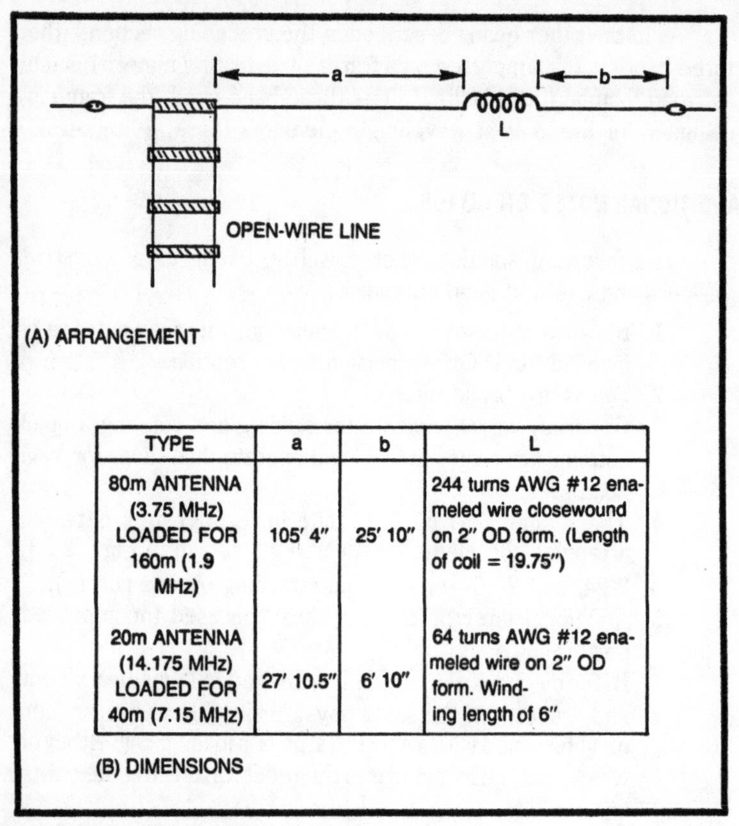

TYPE	a	b	L
80m ANTENNA (3.75 MHz) LOADED FOR 160m (1.9 MHz)	105′ 4″	25′ 10″	244 turns AWG #12 enameled wire closewound on 2″ OD form. (Length of coil = 19.75″)
20m ANTENNA (14.175 MHz) LOADED FOR 40m (7.15 MHz)	27′ 10.5″	6′ 10″	64 turns AWG #12 enameled wire on 2″ OD form. Winding length of 6″

(B) DIMENSIONS

Fig. 2-26. Loaded end-fed half-wave antenna.

Loaded Antenna

When an antenna is too short for a desired operating frequency, its electrical length can be increased by inserting a loading coil. A single loading coil is sufficient in an end-fed antenna (Fig. 2-26A), but two coils—one in each half of the radiator—are required in a center-fed antenna (Fig. 2-27A).

A loading coil contains approximately the additional length of wire needed to make an antenna long enough for the desired frequency, and is wound with wire (usually AWG #12 or #14) heavy enough to handle the antenna current. The coil diameter is kept small enough for mechanical feasibility. Common applications are the loading of an 80-meter antenna for 160-meter operation, and the loading of a 20-meter antenna for 40-meter operation. Conventional antennas for 10 and 20 meters usually can be fitted into most locations and do not need to be loaded up from shorter antennas.

For mechanical and aesthetic reasons, the loading coil might seem best located at the very end of the radiator element. However, in a half-wave antenna, a voltage node is at the very end, and some current must flow through the coil for it to operate. Hence, the coil is inserted instead at a point that is 0.05 wavelength (at the loaded frequency) in from one end.

Follow these steps in designing a loading coil: (1) Record the length of the present antenna as l_1. (2) Using formulas given earlier in this chapter, calculate the length (l_2) that a longer antenna must have for the desired frequency. This step often may be simplified by taking the lengths from the dimension tables given with the drawings in this chapter. (3) Calculate the additional length (l_3) needed:

$$l_3 = l_2 - l_1 \qquad (2\text{-}19)$$

Where each dimension is in feet and inches.

This is the length of wire to be wound into the coil. (4) Calculate the resulting number of turns the coil will have:

$$N = 12l_3/3.14D \qquad (2\text{-}20)$$

where N = number of turns,
l_3 = length of wire in feet, and
D = selected diameter of coil in inches.

(5) Determine the distance d—corresponding to 0.05 wavelength from the free end of the radiator—at which the coil must be inserted:

For an end-fed half-wave antenna:

$$d = 49.1/f \qquad (2\text{-}21)$$

For a center-fed half-wave antenna:

$$d = 46.8/f \qquad (2\text{-}22)$$

in each case, d = distance in feet, and
f = frequency in megahertz.

For the reader's convenience, Fig. 2-26B gives all center-band dimensions and coil data for an 80-meter end-fed antenna loaded for 160-meter operation and a 20-meter end-fed antenna loaded for 40-meter operation. Similarly, Fig. 2-27B gives all dimensions and coil winding data for an 80-meter center-fed antenna loaded for 160-meter operation, and a 20-meter center-fed antenna loaded for 40-meter operation.

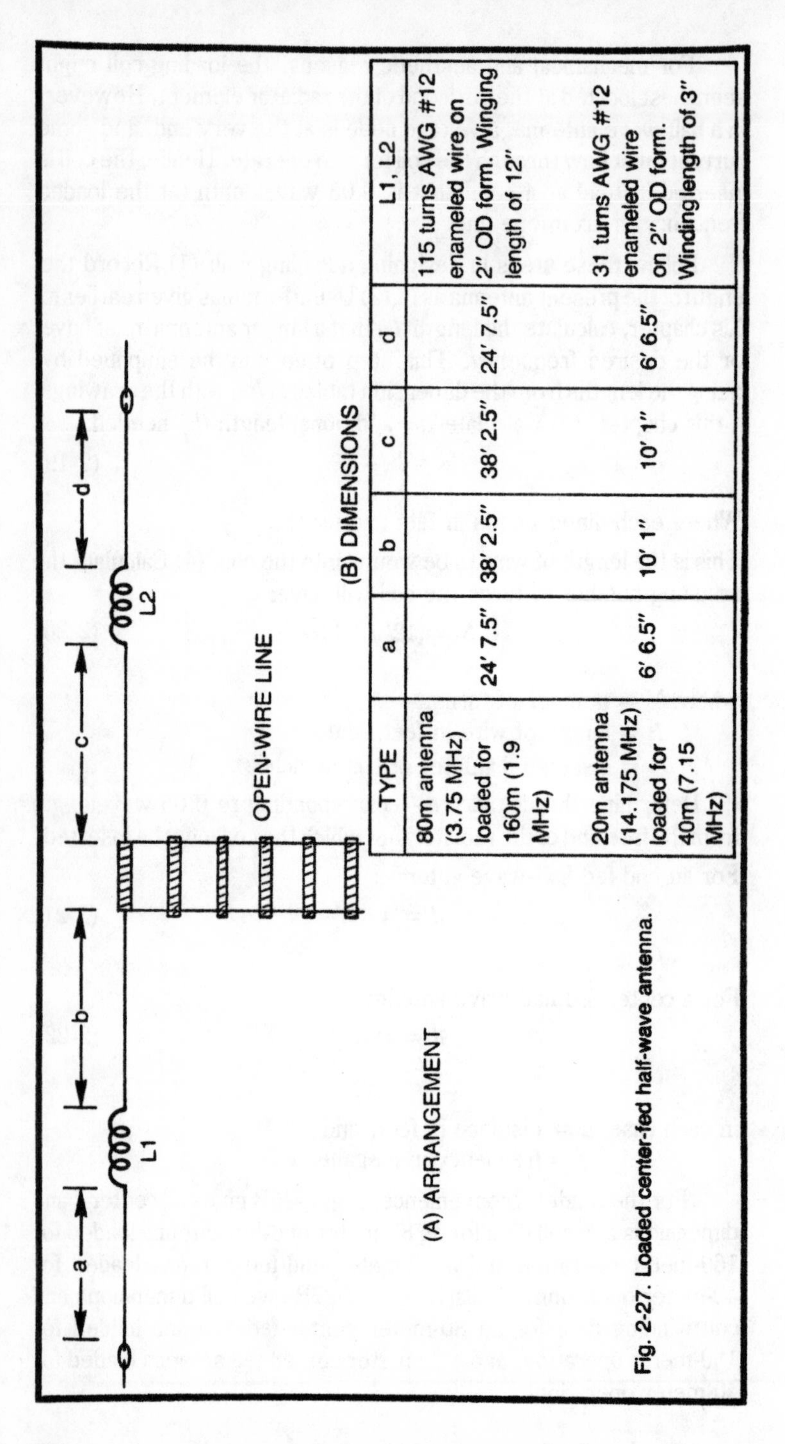

TYPE	a	b	c	d	L1, L2
80m antenna (3.75 MHz) loaded for 160m (1.9 MHz)	24' 7.5"	38' 2.5"	38' 2.5"	24' 7.5"	115 turns AWG #12 enameled wire on 2" OD form. Winging length of 12"
20m antenna (14.175 MHz) loaded for 40m (7.15 MHz)	6' 6.5"	10' 1"	10' 1"	6' 6.5"	31 turns AWG #12 enameled wire on 2" OD form. Winding length of 3"

Fig. 2-27. Loaded center-fed half-wave antenna.

In a vertical antenna, a loading coil is commonly inserted near the top (down approximately 0.05 wavelength at the new operating frequency) to keep radiation resistance high. Such an antenna which is only an eighth-wave long at the desired operating frequency can be loaded to a quarter-wave by means of a suitable coil. This arrangement is shown in Fig. 2-28A. Follow these steps in designing the loading coil: (1) Record the height of the present quarter-wave antenna as h_1. (2) Using Equation 2-1, calculate the height (h_2) that the antenna must have for a quarter-wavelength at the desired frequency. (3) Calculate the additional height (h_3) needed:

$$h_3 = h_2 - h_1 \qquad (2\text{-}23)$$

where each dimension is in feet and inches.

This is the length of wire to be wound into the coil. (4) Calculate the resulting number of turns the coil will have:

$$N = 12h_3/3.14D \qquad (2\text{-}24)$$

where N = number of turns,
h_3 = additional required height in feet, and
D = diameter of coil in inches

(5) Determine the distance d—corresponding to 0.05 wavelength from the top of the antenna—at which the coil must be inserted:

$$d = 46.8/f \qquad (2\text{-}25)$$

where d = distance in feet, and
f = frequency in megahertz.

For the reader's convenience, Fig. 2-28B gives all dimensions and coil winding data for quarter-wave vertical antennas top-loaded for conversion from the center frequency in one band to the center frequency in the next adjacent lower frequency band: 80 meters to 160 meters, 40 meters to 80 meters, and 20 meters to 40 meters.

Loading coils may be used in beam-type antennas, as well as in the types shown in Figs. 2-26, 2-27, and 2-28 (one or more coils are inserted into each radiator, director, and reflector). Also, while those shown in Figs. 2-26 and 2-27 are wire-type antennas, coils are often inserted in antenna elements made of rod or tubing (as in Fig. 2-28—most verticals are tubular—and in most beam antennas).

Loading coils may be wound on either rods or tubes of the best obtainable insulating material. Wood may be used if it is well impregnated to improve its dielectric properties and to safegaurd it against moisture. Also, the coil winding must be protected against

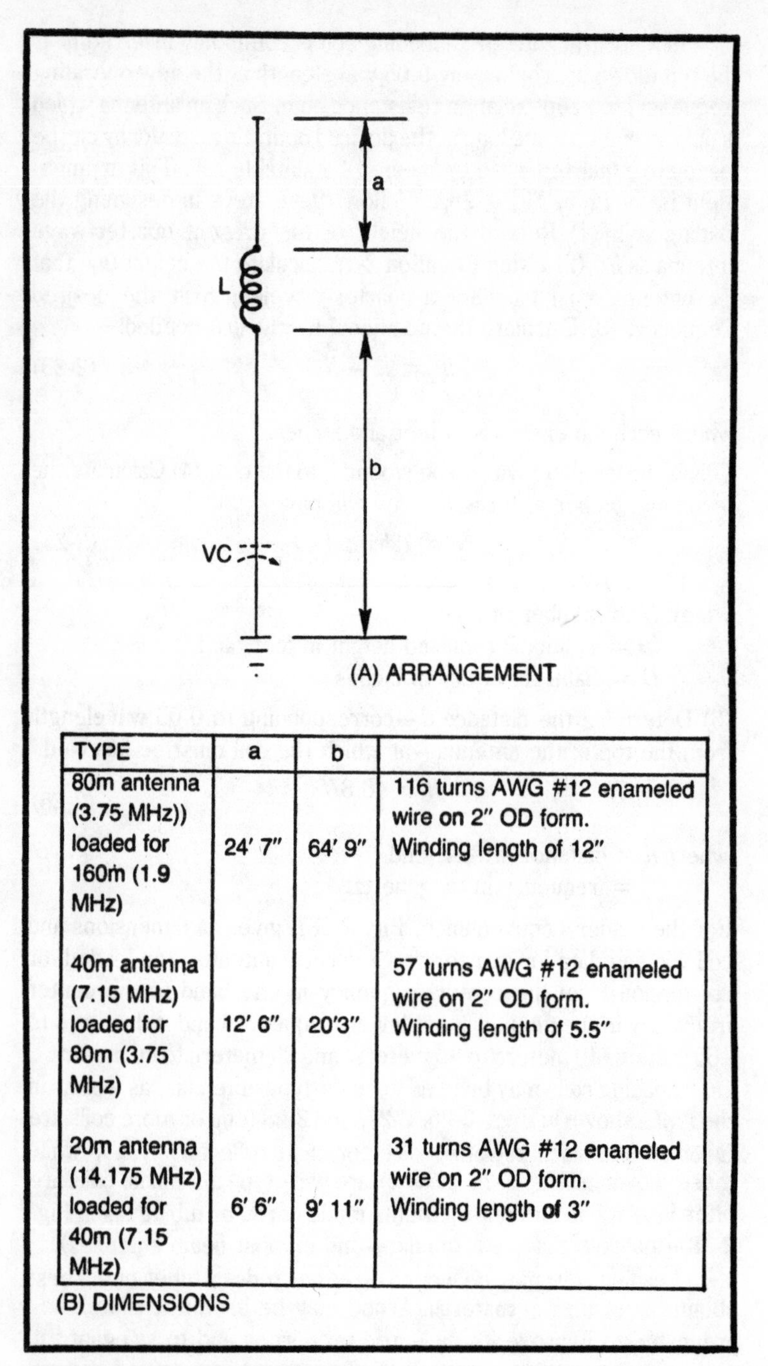

TYPE	a	b	
80m antenna (3.75 MHz)) loaded for 160m (1.9 MHz)	24′ 7″	64′ 9″	116 turns AWG #12 enameled wire on 2″ OD form. Winding length of 12″
40m antenna (7.15 MHz) loaded for 80m (3.75 MHz)	12′ 6″	20′3″	57 turns AWG #12 enameled wire on 2″ OD form. Winding length of 5.5″
20m antenna (14.175 MHz) loaded for 40m (7.15 MHz)	6′ 6″	9′ 11″	31 turns AWG #12 enameled wire on 2″ OD form. Winding length of 3″

(B) DIMENSIONS

Fig. 2-28. Loaded vertical antenna.

the weather. For close adjustment, a loading coil can be provided with taps. However, when a coil is mounted near the top of a vertical antenna, access to it for adjustment may be difficult without repeatedly lowering the antenna. In this instance, close tuning without touching the coil can be obtained with a variable capacitor (*VC* in Fig. 2-28A) inserted as close to ground as practicable. When working with quarter-wave verticals, the reader may find Table 2-2 of value.

Inverted-V Antenna

The *inverted-V* antenna (also called an *inverted dipole*) is a dipole antenna in which the two halves of the radiator are pulled down toward ground at an angle, rather than being horizontal. This arrangement is shown in Fig. 2-29A. The steeper the angle of the inverted-V antenna, the more horizontal space is saved. However, the steeper the angle, the higher the mast must be.

The inverted-V antenna has the advantage that only a single mast is required. The auxiliary supports for the lower ends of the

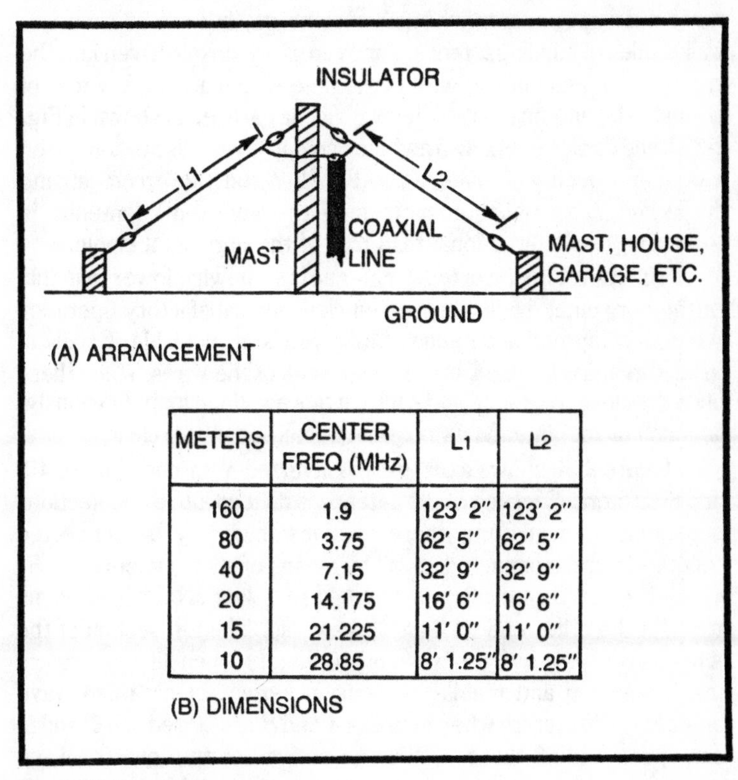

METERS	CENTER FREQ (MHz)	L1	L2
160	1.9	123′ 2″	123′ 2″
80	3.75	62′ 5″	62′ 5″
40	7.15	32′ 9″	32′ 9″
20	14.175	16′ 6″	16′ 6″
15	21.225	11′ 0″	11′ 0″
10	28.85	8′ 1.25″	8′ 1.25″

(B) DIMENSIONS

Fig. 2-29. Inverted-V antenna.

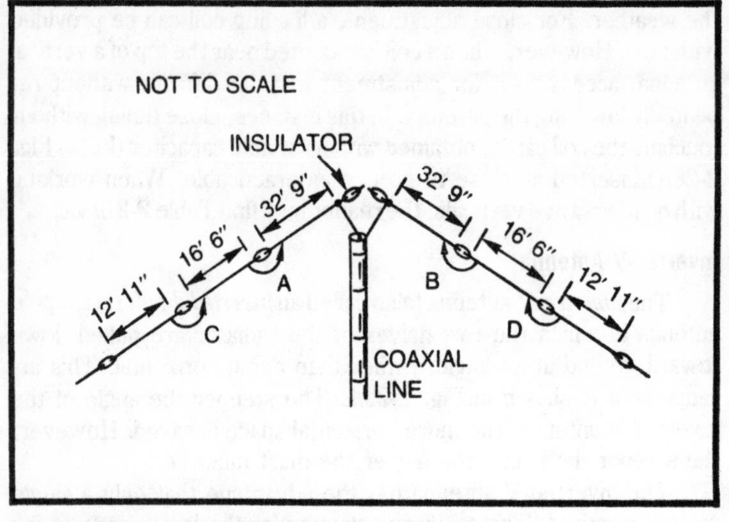

Fig. 2-30. Three-band inverted-V for 20-40-80 meters.

wires need not necessarily be shorter masts, but can be rooftops, walls, sides of buildings, fences, or even short stakes driven into the ground. The ends of the wires can come within a very few feet of ground. The antenna may be fed with a coaxial line, as shown in Fig. 2-29A and earlier in Fig. 2-5 and the accompanying discussion, or by a 600-ohm open-wire line, as in Fig. 2-7 and the accompanying discussion. The radiation pattern of the inverted-V antenna is somewhat less bidirectional than that of the horizontal dipole.

The gain of the inverted-V antenna is somewhat lower than that of the horizontal dipole, but is sufficient for satisfactory operation where a compromise antenna of this type is unavoidable. Excellent insulators must be used at the lower ends of the wires, since these ends are close to ground and each carries a voltage loop. Obviously, as much of the antenna as possible should be in the clear.

Figure 2-30 shows a three-band inverted-V antenna for 20, 40, and 80 meters. Each leg of the antenna is divided into three sections separated by insulators; these sections thus may be connected together and disconnected at will by means of short jumpers (*A*, *B*, *C*, and *D*), each connected at one end to a wire section and terminated at the other end with an alligator clip. The top section of the antenna acts as a quarter-wave on 40 meters when all jumpers are open. The top **and** middle sections in series act as a 3/4-wave antenna on 20 meters when jumpers *A* and *B* are closed and *C* and *D* are open. And all three sections in series act as a quarter-wave antenna on 80 meters when all jumpers are closed.

The dimensions given in Fig. 2-30 are for the center frequency in each of the bands. Similar antennas can be designed for other band combinations, such as 10-20-40 meters and 40-80-160 meters.

Bent Antenna

Very often, horizontal space is restricted at a particular location, but there is ample vertical space for erection of an antenna. Nevertheless, when the desired frequency would mean a prohibitive height, mechanically, for a vertical antenna, some horizontal spread must also be used. In this instance, a horizontal antenna may be bent, part of it being vertical and part of it horizontal. Admittedly, this arrangement will not be as good an antenna as an unbent one, but at some locations it is the only possibility for 40-, 80-, or 160-meter operation.

Figure 2-31A shows an end-fed (zepp) antenna which has been bent. In this arrangement, it is desirable that the horizontal portion (L_2) be as long as practicable, with respect to the vertical portion (L_1);

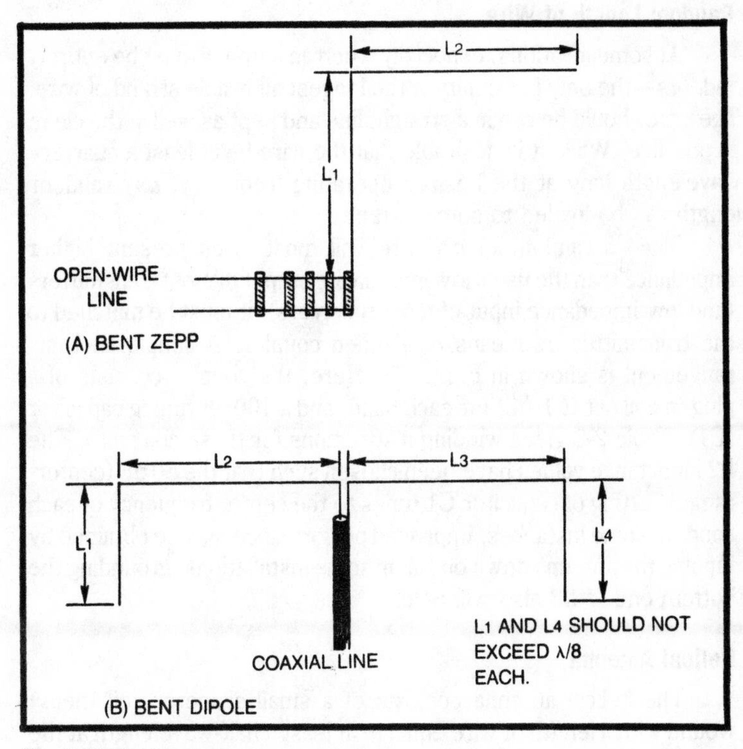

Fig. 2-31. Bent antennas.

however, good performance is obtainable even when $L_2 = L_1$. Figure 2-31B shows a bent dipole. Here, each leg of the radiator is bent vertically down at its end. For best performance, the vertical sections (L_1 and L_4) should not exceed an eighth-wavelength each. This means that ideally L_1, L_2, L_3, and L_4 each is an eighth-wavelength. Fair operation is obtained, however, when L_1 and L_4 are not an eighth-wavelength; but under all circumstances, L_1 should equal L_4.

Most bent antennas require somewhat greater total length than the value calculated for the unbent antenna of the same type. To take care of this requirement, cut the antenna approximately 10 percent longer than the calculated value, then prune it for resonance.

In some antennas of this type, the bent-down section can be attached to the wooden mast; however, this is not advisable with a metal mast. In a great many cases, the bent-down portion will be vertical; nevertheless, it can—where desirable—descend from the antenna at an angle.

Random Length of Wire

At some locations, especially when an antenna must be entirely indoors—the only possibility is the longest allowable strand of wire. The wire should be run in a straight line and kept as well in the clear as possible. While it is desirable that the wire be at least a quarter-wavelength long at the desired operating frequency, *any* random length can be loaded to some extent.

Since a random-length wire will most often present higher impedance than the usual low impedance output of most transmitters (and low-impedance input of most receivers), it must be matched to the transmitter by means of a tuned coupler. A simple such arrangement is shown in Fig. 2-32. Here, the coupler consists of a plug-in coil set (*L*1-*L*2) for each band, and a 100 pF tuning capacitor (C1). Table 2-3 gives winding instructions for these coil pairs. The L2 inductance values have been chosen such that the 50 pF (center-range) setting of capacitor C1 tunes to the center frequency of each band. In some instances, improved performance may be obtained by tapping the antenna down on L2; in some installations, grounding the bottom end of L2 also will help.

Helical Antenna

The helical antenna consists of a small-diameter coil (helix) wound with a length of wire equal to at least a half-wavelength at the desired operating frequency. The coil is wound on a pole or rod of

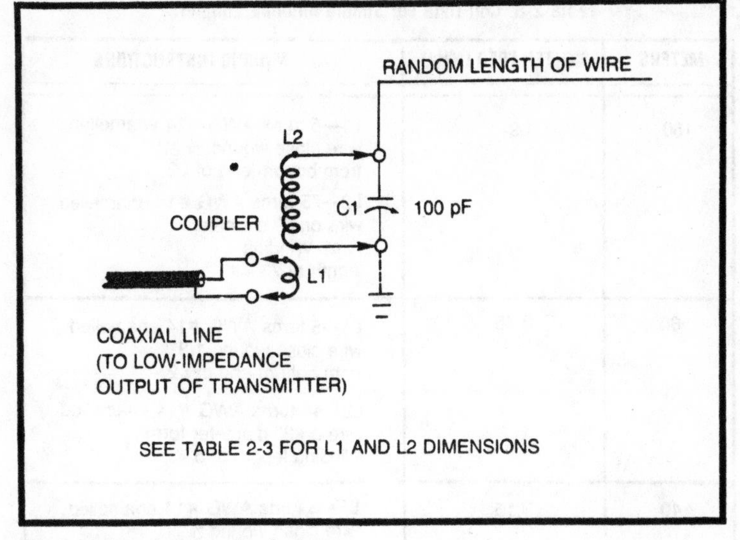

Fig. 2-32. Random length of wire.

weatherproofed insulating material, which is mounted vertically. This antenna behaves electrically as a quarter-wave unit and must be operated against a good ground for best results. It has approximately circular polarization. The helical antenna is suited best for 40-, 80-, and 160-meter restricted-space use, since standard size 10-, 15-, and 20-meter antennas usually present no space problem.

Figure 2-33 shows details. The coil is space wound with AWG #14 enameled wire (double spacing of the wire is adequate). Capacitive loading, needed to reduce the impedance at the end of the antenna, is provided by a 12-inch whip (a piece of AWG #12 wire is sufficiently rigid for this purpose), as shown in Fig. 2-33A, or a 6-inch-diameter horizontal metal disc, as shown in Fig. 2-33B. Whether the whip or the disc is used is a matter of individual choice; in either case, it is connected directly to the top end of the winding. For resonating the antenna at the operating frequency, an adjustable loading coil, *L*, is mounted at the base of the antenna (a suitable variable inductor is Johnson 229-203). This coil must be protected by a weatherproof enclosure.

Figure 2-33C gives data for 160-, 80-, and 40-meter helical antennas. Here, column 3 gives the length of wire corresponding to a half-wavelength at the center-frequency in each band. Column 4 shows the number of turns obtained by double-space winding this length of wire on a 1-inch-diameter rod. Column 5 gives the resulting length of the coil, which in turn determines the length of the rod. It is

Table 2-3. Coil Data For Simple Antenna Coupler.

METERS	CENTER FREQ (MHz)	WINDING INSTRUCTIONS
160	1.9	L1—5 turns AWG #24 enamelled wire close wound ⅛" from bottom end of L2 L2—75 turns AWG #14 enamelled wire on 3" diameter form. Winding length of 7"
80	3.75	L1—5 turns AWG #14 enamelled wire close wound 1/8" from bottom end of L2 L2—48 turns AWG #14 enamelled wire on 2" diameter form. Winding length of 5"
40	7.15	L1—3 turns AWG #14 enamelled wire close wound ⅛" from bottom end of L2 L2—20 turns AWG #14 enamelled wire on 2" diameter form. Winding length of 3"
20	14.175	L1—2 turns AWG #14 enamelled wire close wound ⅛" from bottom end of L L2—9 turns AWG #14 enamelled wire on 2" diameter form. Winding length of 2"
15	21.225	L1—2 turns AWG #14 enamelled wire close wound ⅛" from bottom end of L2 L2—5 turns AWG #14 enamelled wire on 2" diameter form. Winding length of 1.5"
10	28.85	L1—1 turn AWG #14 enamelled wire ⅛" from bottom end of L2 L2—3 turns AWG #14 enamelled wire on 2" diameter form. Winding length of 1"

advisable to leave a space of at least 2 inches bare at the bottom of the rod to accommodate mounting hardware. Thus, a 12-foot pole will be entirely adequate for the 160-meter antenna, a 6-foot one for the 80-meter antenna, and a 3-foot one for the 40-meter antenna.

As stated above, the values in Fig. 2-33C are based upon a diameter of 1 inch. For a different diameter (D_x), multiply the figures in columns 4 and 5 by $1/D_x$. Thus, for a half-inch-diameter rod, the number of turns in the 160-meter antenna will be 941 (1/0.5) = 941(2) = 1882; and the length of the winding is 120(1/0.5) = 120(2) = 240 inches = 20 feet.

For suitable grounds for use with the helical antenna, see the section entitled *Ground Connection for Marconi Antenna*, located near the beginning of this chapter.

Inclined (Tilted) Dipole

As a compromise unit where horizontal space is limited, a dipole antenna may be tilted, as shown in Fig. 2-34A. The insulated lower end of the antenna can be fastened to a low building, the rooftop, a fence, a stake driven into the ground, or a similar convenient point of attachment. Good results have been reported with the lower end of the wire only a few inches from ground; however, superior insulation is required at the lower end of the inclined dipole since this is the location of a voltage loop; the closer this point is to ground, the better the insulation must be. Maximum radiation is in the direction of the arrow in Fig. 2-34A. This is somewhat of a low angle. Design data for the dipole itself (which is standard in every way, except that it is tilted) are given in the section entitled *Simple Dipole* and Fig. 2-5.

Figure 2-34B shows height H that is required for various common desired horizontal distances D for 20-, 40-, 80-, and 160-meter inclined dipoles. Values are given, as shown, for the center frequency in each band. From this table, it is seen that the height required at the lower frequencies becomes prohibitive except in some special cases (for example, the single mast for a 160-meter inclined antenna confined to a 30-foot space must be 244 feet 6 inches high).

If it is desired to know the angle x between the wire and ground, this may be found from the relationship: $x = \arctan H/D$. Thus, for the 20-meter antenna erected in a 30-foot horizontal space, $D = 30$ feet, and $H = 13$ feet 9 inches = 13.75 feet; and $x = \arctan 13.75/30 = \arctan 0.458333 = 24.62°$.

MOBILE ANTENNAS

Rigid vertical ham antennas for installation on automobiles are commercially available in numerous makes and models, and most hams undoubtedly will prefer to buy one of these ready-made units. A few users, however, may choose to build their own, especially in

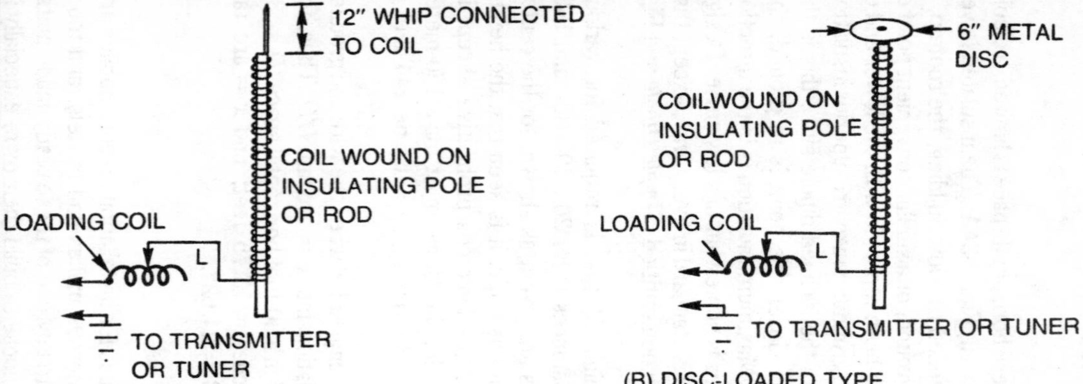

12" WHIP CONNECTED TO COIL
COIL WOUND ON INSULATING POLE OR ROD
LOADING COIL
L
TO TRANSMITTER OR TUNER

(A) WHIP-LOADED TYPE

6" METAL DISC
COILWOUND ON INSULATING POLE OR ROD
LOADING COIL
L
TO TRANSMITTER OR TUNER

(B) DISC-LOADED TYPE

METERS	CENTER FREQ (MHz)	WIRE LENGTH	APPROX. TURNS * (1" DIA.)	APPROX. WINDING LENGTH
160	1.9	246' 4"	941	10' 0"
80	3.75	124' 10"	477	5' 1"
40	7.15	65' 6"	250	2' 8"

*—AWG #14 ENAMELED WIRE DOUBLE SPACED

(C) DIMENSIONS

Fig. 2-33. Helical antenna.

an emergency, and the following information is offered for their benefit.

The following pointers are worth noting: The antenna must be mounted as high as possible on the vehicle. This requirement favors the rear deck (trunk lid) instead of the rear bumper, although the latter is often used. The car top would be an even better location, but the length of the antenna prohibits use of the top except at those very high frequencies at which the antenna is quite short. The antenna must be provided with a suitable mount which will provide sturdy support, insulate the antenna from the car, and allow feed-through clearance for the coaxial line connecting the antenna to the radio unit. At least the upper portion of the antenna should be telescoping for close adjustment.

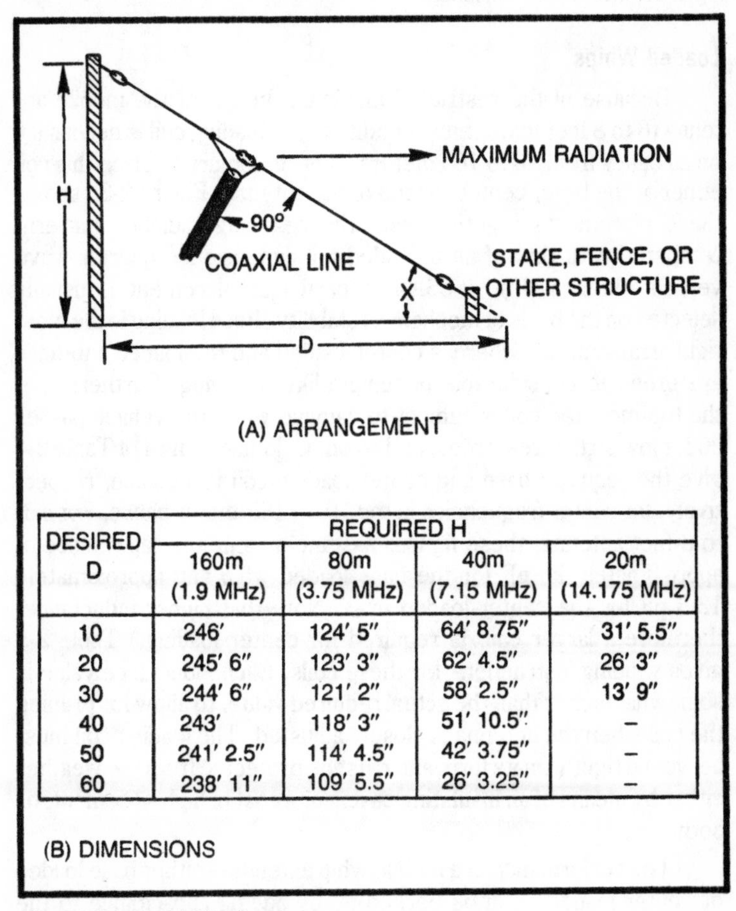

DESIRED D	REQUIRED H			
	160m (1.9 MHz)	80m (3.75 MHz)	40m (7.15 MHz)	20m (14.175 MHz)
10	246′	124′ 5″	64′ 8.75″	31′ 5.5″
20	245′ 6″	123′ 3″	62′ 4.5″	26′ 3″
30	244′ 6″	121′ 2″	58′ 2.5″	13′ 9″
40	243′	118′ 3″	51′ 10.5″	–
50	241′ 2.5″	114′ 4.5″	42′ 3.75″	–
60	238′ 11″	109′ 5.5″	26′ 3.25″	–

(B) DIMENSIONS

Fig. 2-34. Inclined (tilted) antenna.

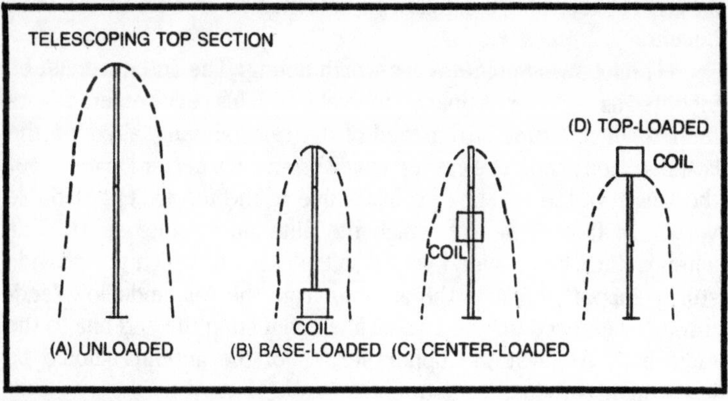

Fig. 2-35. Mobile whip antennas.

Loaded Whips

Because of the restricted maximum height of the mobile antenna (6 to 8 feet under most conditions), a loading coil is necessary on all bands from 10 to 160 meters. It is customary to place this coil either at the base, center, or top of the antenna. Figure 2-35 shows these placements together with the resulting radiation pattern, compared with that of an unloaded full-dimensioned quarter-wave vertical antenna (Fig. 2-35A). A particular placement is usually selected on the basis of mechanical stability. Base loading is the most rigid arrangement, whereas center-loaded and top-loaded antennas are prone to considerable pendulum-like swinging. Furthermore, the top-mounted coil is subject to damage when the vehicle passes under low structures, trees, and so on. Columns 2 and 4 of Table 2-4 give the required base and center loading-coil inductance, respectively, for center frequencies in the 10- to 160-meter bands. For a 6- to 8-foot antenna, these figures assume an antenna capacitance of approximately 25 pF for the base-loaded whip and approximately 12.5 pF for the center-loaded one. (Note that higher inductance, therefore a larger coil, is required for center loading.) Table 2-5 gives winding instructions for these coils. Each inductance value is somewhat higher than the actual required value, to allow for pruning the coil when the antenna is closely adjusted. The loading coil must be wound tightly on its form and must be protected from the weather either by means of an insulating casing or a coat of high-Q cement, or both.

The performance of a mobile whip antenna—either base loaded or center loaded—can be perked up by adding capacitance to the portion of the antenna above the loading coil. This capacitance tends

Table 2-4. Mobile Antenna Data.

METERS	CENTER FREQUENCY (MHz)	LOADING-COIL INDUCTANCE*	
		BASE LOADING (μH)	CENTER LOADING (μH)
160	1.9	285 (L1)	561 (L7)
80	3.75	72 (L2)	144 (L8)
40	7.15	19.8 (L3)	39.6 (L9)
20	14.175	5.0 (L4)	1.0 (L10)
15	21.225	2.25 (L5)	4.49 (L11)
10	28.85	1.22 (L6)	2.43 (L12)

* Table 2-5 for coil winding instructions

to resonate with the inductance of the coil. Figure 2-36 shows several simple ways of adding capacitance. In Fig. 2-36A, a single radial (stiff, horizontal wire or rod) is attached to the antenna and may be bent up and down to vary the capacitance. The length of the radial may be pruned during adjustment of the antenna. In Fig. 2-36B, several radials are attached for increased capacitance. In Fig. 2-36C, the radials are connected together by an outer ring for still higher capacitance and form a capacitance hat or capacitance wheel. In Fig. 2-36D, the largest amount of capacitance is obtained by using a disc of sheet metal or metal screen. Whichever capacitance device is employed, it must be connected to the radiator right at the top of the loading coil.

Since the base impedance of the whip antenna is notably lower than that of the coaxial line that brings power from the transmitter, an impedance-matching network is required. Figure 2-37 shows use for this purpose of a popular, simple L-network, so called from its

Table 2-5. Mobile Loading Coil Data.

COIL	INDUCTANCE (μH)	NO. TURNS	AWG	DIAMETER (IN.)	WINDING LENGTH (IN.)
L_1	285	85	20	3	4
L_2	72	40	20	3	3
L_3	20	30	18	2	2
L_4	5	15	12	2	2
L_5	2.5	10	12	1.5	1.5
L_6	1.25	7	12	1.5	1.5
L_7	560	118	20	3	9
L_8	144	66	20	3	5
L_9	40	35	18	2	2
L_{10}	1	6	12	1.5	1.5
L_{11}	4.5	14	12	2	2
L_{12}	2.5	10	12	1.5	1.5

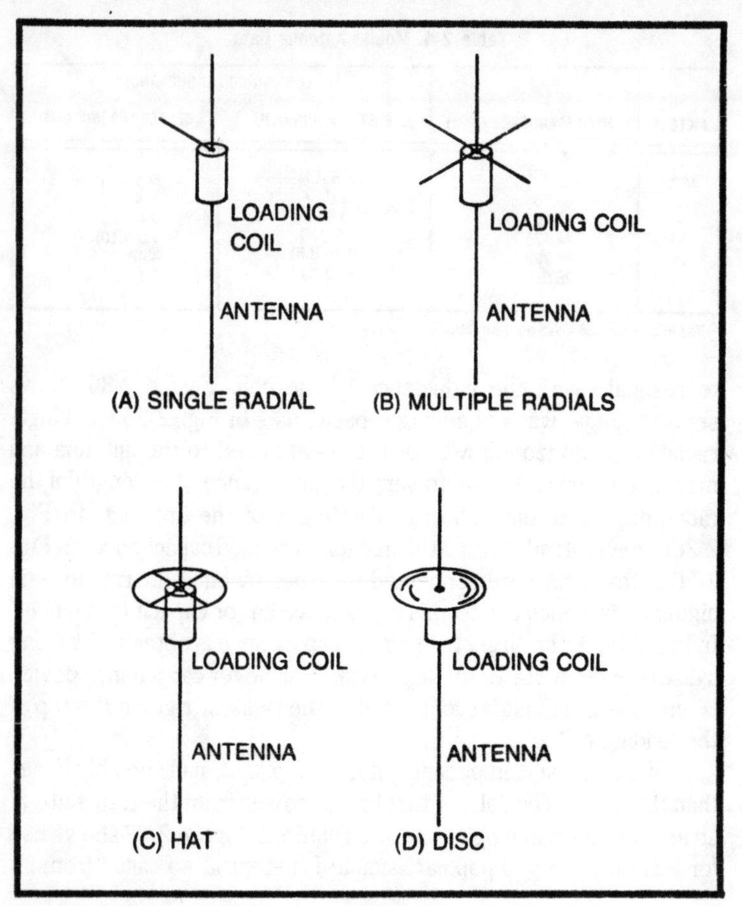

Fig. 2-36. Adding capacitance to mobile whip.

resemblance to an inverted letter *L* when the inductor is drawn horizontal in the diagram. Table 2-6 gives inductance *L* and capacitance *C* values for the L-network, at the center frequency in the 10- to 160-meter bands. These values are based upon an antenna capacitance of approximately 12.5 pF for the base-loaded whip, and an average antenna base impedance of 22 ohms ($C_{pF} = 3917.2/f_{MHz}$, and $L_{\mu H} = 3.8216/f_{MHz}$). In practice, the inductor sometimes is dispensed with and the required *L* (from Table 2-6) is obtained by adding turns to the base or center loading coil. For close adjustment of the L-network, a part of the required capacitance, *C*, is made variable. Thus, for the 184 pF capacitance shown for 15 meters in Table 2-6, a 100 pF air-trimmer-type midget variable capacitor may be connected in parallel with a 120 pF fixed capacitor, and the

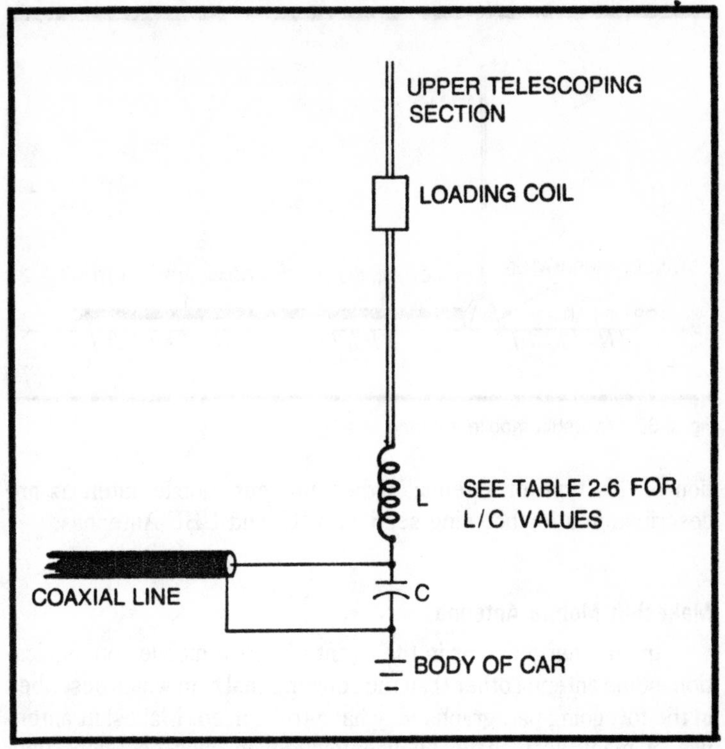

Fig. 2-37. L-network for coupling to whip.

required 184 pF will be obtained when the variable unit is set just slightly beyond midcapacitance. This will allow ample tuning range.

Additional Mobile Antennas

Helical whip antennas are constructed and used by some hams. See the section entitled *Helical Antenna* in this chapter for a descrip-

Table 2-6. L-Network Constants.

METERS	CENTER FREQUENCY (MHz)	L (μH)	C (μH)
160	1.9	2	2062
80	3.75	1	1044
40	7.15	0.53	548
20	14.175	0.27	276
15	21.225	0.18	184
10	28.85	0.13	136

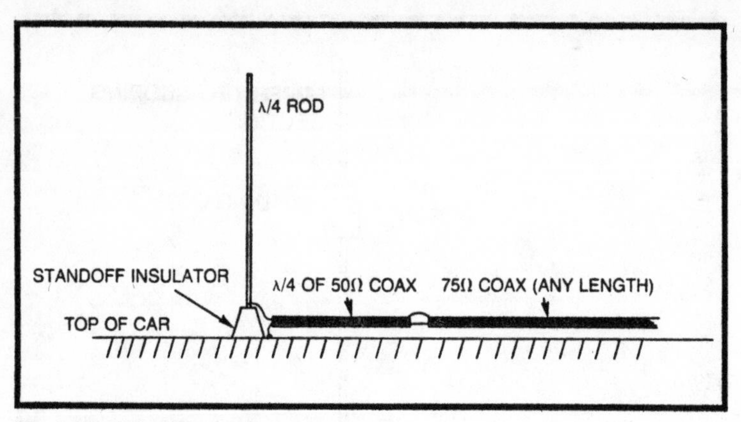

Fig. 2-38. Makeshift mobile antenna.

tion of this type of antenna. Other amateur mobile antennas are described in a forthcoming section, VHF and UHF Antennas.

Makeshift Mobile Antennas

In an emergency or in the event of a rare mobile communication, some antenna other than the conventional ham whips described in the foregoing paragraphs may have to be used. Makeshift antennas of many sorts have been employed on such occasions with various degrees of success.

On wavelengths of 20 meters and lower (band center frequencies of 14.175 MHz and higher), where a half-wavelength is 16.5 feet or shorter, a quarter-wavelength rod or pipe may be stood temporarily on a standoff insulator on top of the car, as shown in Fig. 2-38. (The insulator may be attached permanently to the roof, and the antenna fastened to it only when needed.) For impedance matching at the base of this antenna, a quarter-wavelength of 50-ohm coaxial cable, such as RG58, is used. And this section is connected by the shortest practicable leads to a 75-ohm coaxial line (such as RG59) which may be run any length to the transmitter and receiver. The metal body of the car serves as a ground plane. This scheme suffers mostly from its limitation to standing-car operation; except for the shortest antennas, passing under obstructions when the car is in motion will result in bending or grounding the antenna. Another sometimes undesirable feature is the permanent attachment of the standing insulator for only occasional use.

In an emergency, it is possible sometimes to use the regular whip radio antenna that is already installed in the car. Since this is a telescoping antenna, it can, for some frequencies, be adjusted to a half-wavelength. The experimenter must keep in mind, however, that the insulation between these broadcast-receiver antennas and the car body is not usually of the low-loss type; and this can introduce leakage, especially when the mobile transmitter has appreciable power output.

VHF AND UHF ANTENNAS

Many of the antenna types described earlier in this chapter can be used also at frequencies of 50 MHz and higher, provided that the elements are properly scaled down in size. For this purpose, Table 2-7 gives the physical lengths in inches of various wavelengths at the center frequency of the 1-, 2-, and 6-meter bands and of the 420 MHz band.

Adapting Lower-Band Types

The following schedule explains how to convert various popular types of antenna to VHF models:

Simple Dipole. Refer to Fig. 2-5A and the accompanying text. Use Table 2-7 to calculate dimensions.

Center-Fed Hertz. Refer to the section earlier in this chapter covering this type of antenna and the appropriate subsection (*With Open-Wire Resonant Feeders* etc.) and use Table 2-7 to calculate dimensions.

Coaxial Antenna. Refer to the section covering this type of antenna appearing earlier in this chapter and Figs. 2-15A and 2-15B. Use Table 2-7 to calculate dimensions (dimensions for 2 and 6 meters are already given in Fig. 2-15C).

Ground-Plane Antenna. Refer to the section earlier in this chapter covering this type of antenna and Figs. 2-16A and 2-16B. Use Table 2-7 to calculate dimensions (dimensions for 2 and 6 meters are already given in Fig. 2-16C).

Table 2-7. VHF/UHF Antenna Dimensions.

METERS	CENTER FREQ (MHz)	LENGTH (in.)					
		λ	5λ/8	λ/2	λ/4	λ/5	0.15λ
6	52	216*	135*	108***	54	43.5	32
2	146	77÷	48	38	19.25	15.375	11
1	222.5	50.5	31.5	25.25	12	10	7
0.70	435	29	16	13	6	5.1875	3.875

*—18′ **—11′ 3″ ***—9′ ÷—6′ 5″

Long-Wire Antenna. Refer to the section earlier in this chapter covering this type of antenna and Fig. 2-17. Use Table 2-7 to calculate dimensions (dimensions for 6 meters are already given in Fig. 2-17B).

V-Antenna. Refer to the section earlier in this chapter covering this type of antenna and Figs. 2-18A and 2-18B. Use Table 2-7 to calculate dimensions (dimensions for 6 meters are already given in Fig. 2-18C).

Rhombic Antenna. Refer to the section earlier in this chapter covering this type of antenna and Figs. 2-19A and 2-19B. Use Table 2-7 to calculate dimensions (dimensions for 6 meters are already given in Fig. 2-19C).

Two-Element Beams. Refer to the section earlier in this chapter covering this type of antenna and Figs. 2-20A, 2-20B, and 2-20C. Use Table 2-7 to calculate dimensions (dimensions for 6 meters are already given in 2-20D).

Three-Element Beams. Refer to the section earlier in this chapter covering this type of antenna and Figs. 2-21A and 2-21B. Use Table 2-7 to calculate dimensions (dimensions for 6 meters are already given in Fig. 2-21D).

IMPORTANT

WHEN ONE OF THE FOREGOING ANTENNAS IS MOUNTED VERTICALLY, RUN THE TRANSMISSION LINE AWAY FROM ANTENNA HORIZONTALLY FOR AT LEAST A HALF-WAVELENGTH.

Special Directive Antennas

The greatest attraction of the VHF/UHF region to the antenna experimenter is the opportunity this part of the frequency spectrum affords for working with small-sized, easily rotatable, multielement beam antennas (at lower frequencies, the size of such antennas would be prohibitive). Thus, at VHF and UHF, a large number of short directors and reflectors can be mounted in a comparatively small space, and correspondingly high gain and narrow directivity accordingly obtained. Several beam antennas of this type are described in the following paragraphs. In most of the units shown, dimensions are given in inches according to the formula $\lambda/2 = 5600/f_{MHz}$. In the Yagi arrays, the length of the radiator (driven element) is equal to $5600/f$, the length of the reflector = $5880/f$, and the length of the director = $5320/f$ (this means that the reflector is 5 percent longer than the radiator, and the director is 5 percent shorter than the radiator).

Figure 2-39 shows the layout of a five-element Yagi beam designed for 52 MHz, the center frequency of the 6-meter band. This array consists of the radiator (driven element), one reflector, and three directors. The elements are cut from quarter-inch or half-inch OD aluminum tubing or quarter-inch rod, all of which are supported at their centers. The spacing of the elements (see Fig.

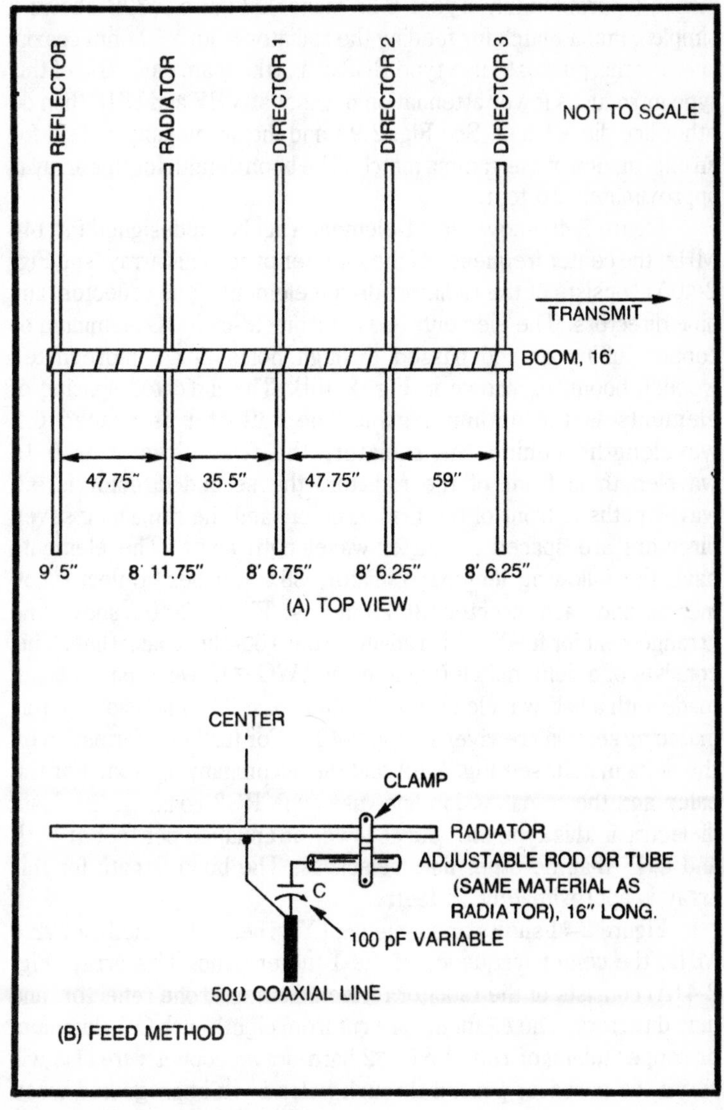

Fig. 2-39. Five-element 52 MHz beam.

2-39A) is for optimum gain: The reflector is spaced 0.2 wavelength behind the radiator, the first director is 0.15 wavelength in front of the radiator, the second director 0.2 wavelength in front of the first director, and the third director 0.25 wavelength in front of the second director. The elements have the following lengths: radiator, 107.7 inches (approximately 8 feet 11.75 inches); reflector, 113 inches (approximately 9 feet 5 inches); and each director 102.3 inches (approximately 8 feet 6.25 inches). Figure 2-39B shows a simple gamma match for feeding the radiator from a 50-ohm coaxial line; for this purpose, use type RG8 coax (the foam dielectric of this type introduces lower attenuation per foot at VHF and UHF than do other line dielectrics). See Fig. 2-9A and the accompanying text for an explanation of the gamma match. The boom length for this array is approximately 16 feet.

Figure 2-40 shows an 11-element Yagi beam designed for 146 MHz, the center frequency of the 2-meter band. This array (see Fig. 2-40A) consists of the radiator (driven element), one reflector, and nine directors. The elements are cut from 18-inch OD aluminum or copper tubing or rod passed through holes in an impregnated wooden boom, as shown in Fig. 2-40B. The indicated spacing of elements is for optimum gain: The reflector is spaced 0.2 wavelengths behind the radiator, the first director is 0.15 wavelength in front of the radiator, the second director is 0.2 wavelengths in front of the first director, and the remaining seven directors are spaced a quarter-wavelength apart. The elements have the following lengths: radiator, 38.25 inches; reflector, 40 inches; and each director, 36.25 inches. Figure 2-40C shows the arrangement for feeding the radiator from a 50-ohm coaxial line. This consists of a delta match (made from AWG #12 wire) plus a balun made with a half-wavelength of 50-ohm coax. All dimensions of this matching section are given in Fig. 2-40C. For further information on the delta match, see Fig. 2-8A and the accompanying text. For the balun and the transmission line, use type RG8 coax, as the foam dielectric in this type introduces lower attenuation per foot at VHF and UHF than do other line dielectrics. The boom length for this array is approximately 15 feet.

Figure 2-41 shows an 11-element Yagi beam designed for 222.5 MHz, the center frequency of the 1-meter band. This array (Fig. 2-41A) consists of the radiator (driven element), one reflector, and nine directors. The elements are cut from eighth-inch OD aluminum or copper tubing or rod (AWG #2 hard-drawn copper wire also will serve the purpose) passed through holes in an impregnated wood boom, as shown in Fig. 2-41B. An all-metal assembly also can be

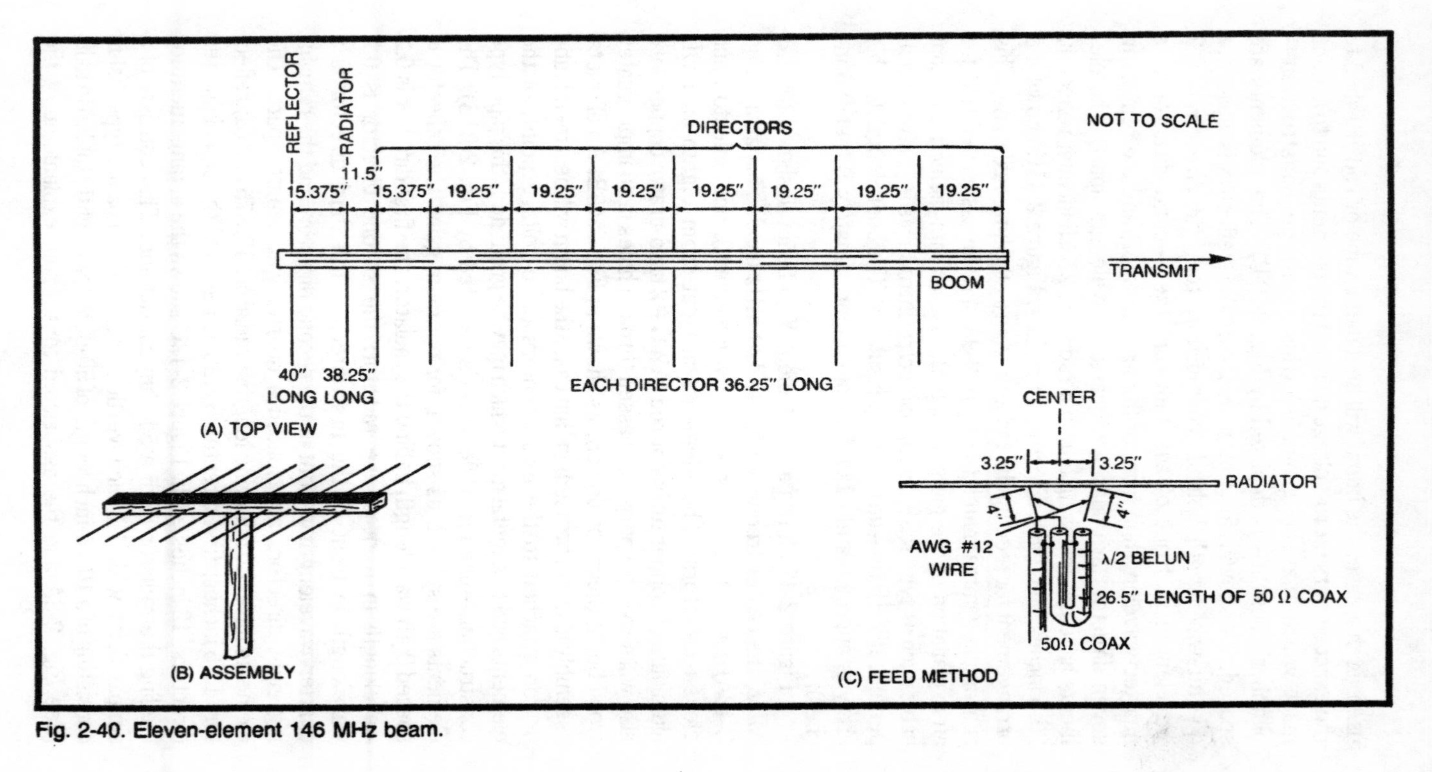

Fig. 2-40. Eleven-element 146 MHz beam.

used; in that case, the boom will be a metal tube or rod attached to the exact centers (zero-voltage point) of the elements and to a metal mast which may be grounded (this pipe-type construction—plumber's delight—is depicted by Fig. 2-21B). The elements are spaced, as shown, for optimum gain: The reflector is spaced a quarter-wavelength behind the radiator, the first director is 0.2 wavelength in front of the radiator, the second director is a quarter-wavelength in front of the first director, and the remaining seven directors are spaced a quarter-wavelength apart. The elements have the following lengths: radiator, 25.25 inches; reflector, 26.5 inches; and each director, 24 inches. Figure 2-41C shows the arrangement for feeding the radiator from a 50-ohm coaxial line. This is a simple gamma match (see Fig. 2-9A and the associated text for an explanation of the gamma match). Use a foam-dielectric coaxial cable such as type RG8, as this dielectric introduces less attenuation per foot at VHF than do other line insulants. The boom length for this array is approximately 117.5 inches (approximately 9 feet 9.5 inches).

Figure 2-42 shows a 13-element Yagi beam designed for 435 MHz, the center frequency of the 420 MHz band. This array (Fig. 2-42A) consists of the radiator (driven element), one reflector, and eleven directors. The elements are cut from eighth-inch OD aluminum or copper tubing or rod (AWG #2 hard-drawn copper wire also will serve the purpose) passed through holes in an impregnated wood boom (beam or dowel), as shown in Fig. 2-42B. An all-metal assembly also can be used; in that case, the boom will be a metal tube or rod attached to the exact centers (zero-voltage point) of the elements and to a metal mast which may be grounded (this pipe-type construction—plumber's delight—is depicted by Fig. 2-21B). The elements are spaced, as shown, for optimum gain: The reflector is spaced 0.15 wavelength behind the radiator, the first director is 0.2 wavelength in front of the radiator; the second director is 0.2 wavelength in front of the first director, the third director is a quarter-wavelength in front of the second director, and the remaining eight directors are spaced a quarter-wavelength apart. The elements have the following lengths: radiator, 12.875 inches; reflector, 13.5 inches; first director, 12.25 inches; and the remaining ten directors, 12 inches each. Figure 2-42C shows the arrangement for feeding the radiator from a 50-ohm coaxial line. This consists of a simple delta with 1.5-inch vertical legs and a coaxial-type balun consisting of a 10.25-inch length of 50-ohm coax tightly folded in half. (See Fig. 2-8A and the associated text for an explanation of the

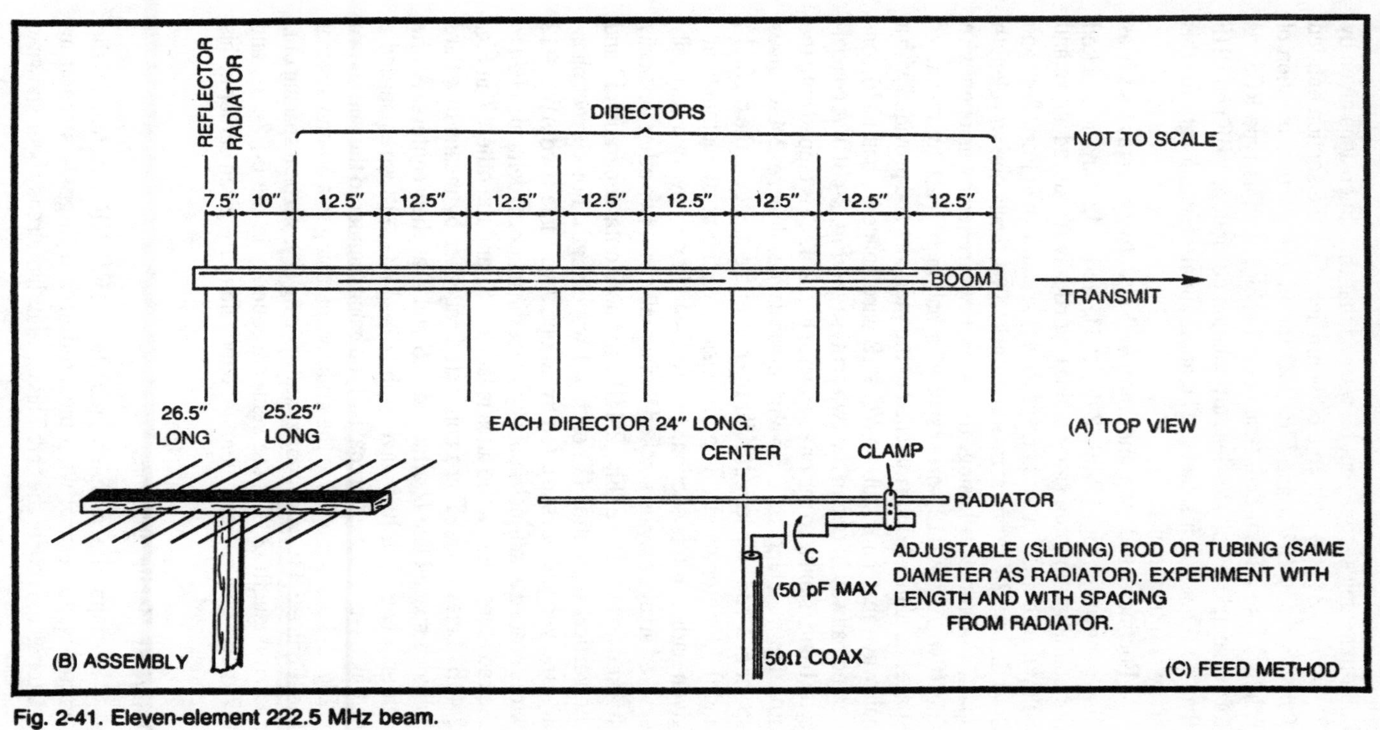

Fig. 2-41. Eleven-element 222.5 MHz beam.

delta-matching section, but be warned that the dimensions given by Equation 2-3 apply only to a 600-ohm line. See the section entitled *Three-Element Beam* and Fig. 2-22 for an introductory mention of baluns.) Use a foam-dielectric coaxial cable, such as type RG8, as this dielectric introduces less attenuation per foot at VHF and UHF than do other line insulants. The approximate boom length for this array is 6 feet 1 inch.

For increased gain and improved directivity, arrays such as those shown in Figs. 2-39, 2-40, 2-41, and 2-42 may be stacked horizontally (mounted side by side), vertically (mounted above and below), or both. Stacked arrays (bays) are shown in Figs. 2-43 and 2-44. In Fig. 2-43, two arrays are stacked vertically one wavelength apart. For correct phasing, the two arrays are fed by an open-wire line (two AWG #12 wires spaced a half-inch apart by means of spacer insulators) delta matched to the radiators and provided with a half-wave matching stub (AWG #12 wire spaced a half-inch) connected halfway between the two arrays (see Fig. 2-43B). A coaxial-type balun (similar to the one shown in Fig. 2-42C) is tapped onto the stub for matching to a 50-ohm coaxial line. Figure 2-43C gives spacing *X* and stub length *Y* for the center frequency in the 6-, 2-, 1-, and 0.70-meter bands. The dimensions of the array elements are given earlier in this section. Figure 2-44 shows how two vertically stacked arrays may be stacked horizontally with two other vertically stacked arrays. Here (Fig. 2-44B), the horizontal separation *Y*) and the vertical separation (*X*) each is 1 wavelength. For correct phasing, the vertical bays are fed by an open-wire line (two AWG #12 wires spaced a half-inch apart by spacer insulators) delta matched to the radiators. One of the lines is labelled *S*, and the other, *T* in Fig. 2-44B. Lines *S* and *T* are connected together by means of a third half-inch-spaced line *U* attached to *S* and *T* at their centers. At the center lf line *U*, a half-wave stub (*Z*)—AWG #12 wires spaced a half-inch—is attached; a coaxial-tyep balun (similar to the one shown in Fig. 2-42C) is tapped to the stub for matching to a 50-ohm coaxial line. Figure 2-44C gives horizontal spacing (*Y*), vertical spacing (*X*), and stub length (*Z*) for the center frequency in the 6-, 2-, 1-, and 0.7-meter bands. Array element dimensions are given earlier in this section.

VHF/UHF Mobile Antennas

Perhaps the simplest mobile VHF/UHF antenna is the quarter-wave vertical whip mounted on a high-grade standoff insulator on the roof of the car. This antenna has already been shown in

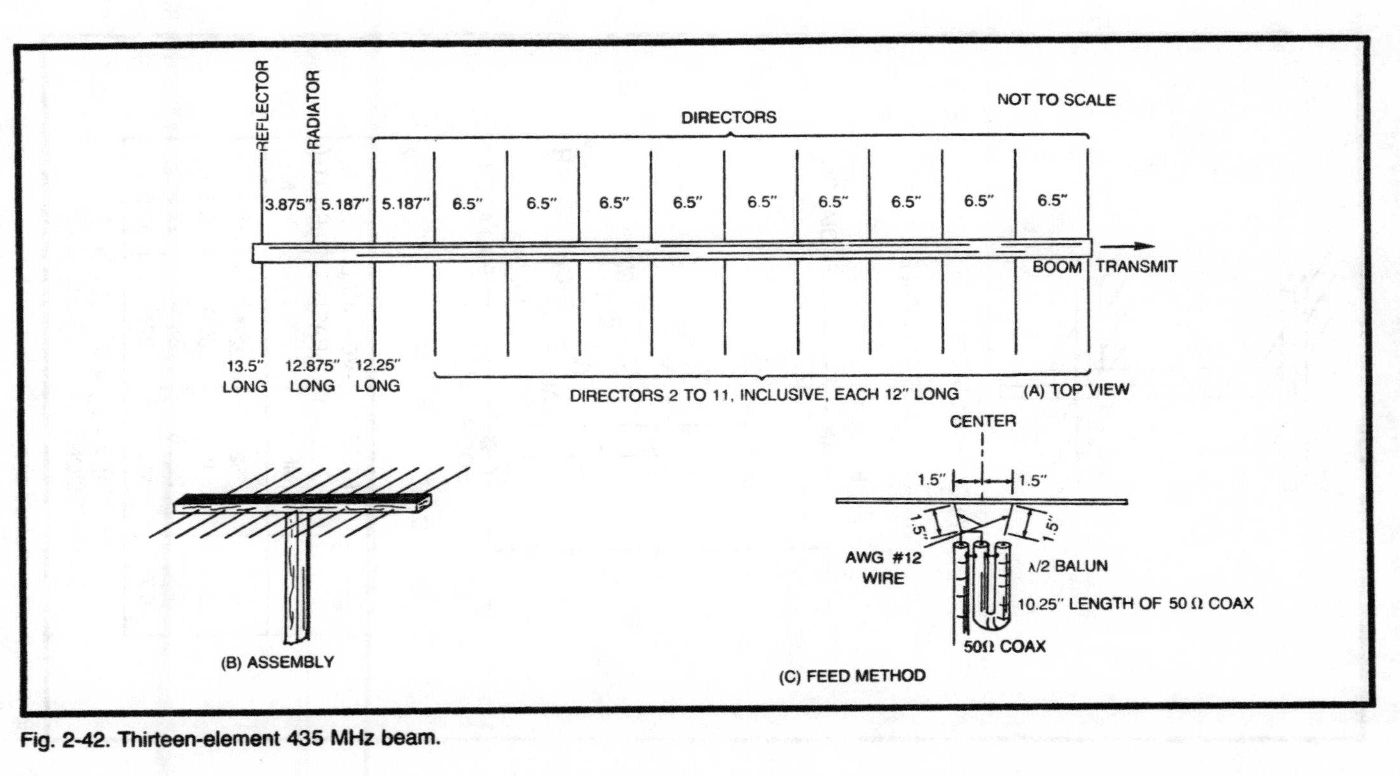

Fig. 2-42. Thirteen-element 435 MHz beam.

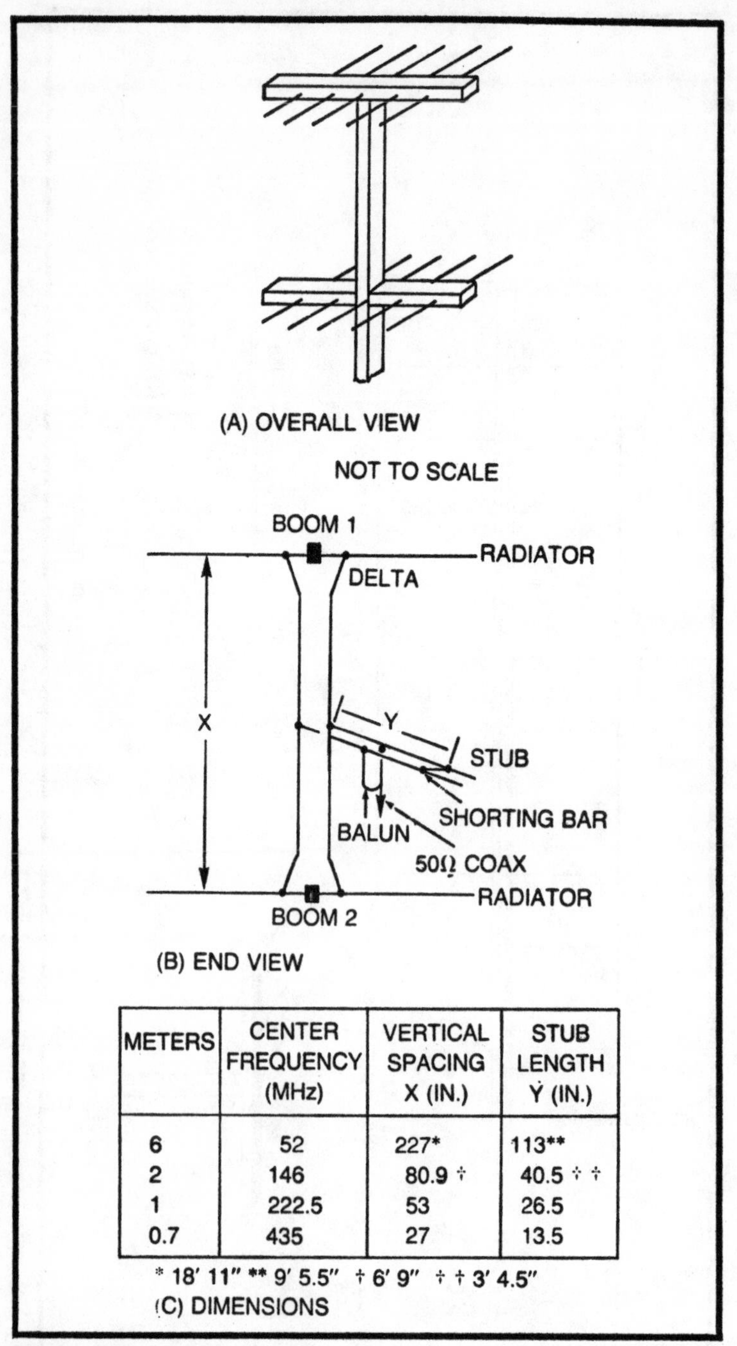

METERS	CENTER FREQUENCY (MHz)	VERTICAL SPACING X (IN.)	STUB LENGTH Y (IN.)
6	52	227*	113**
2	146	80.9 †	40.5 ††
1	222.5	53	26.5
0.7	435	27	13.5

* 18′ 11″ ** 9′ 5.5″ † 6′ 9″ †† 3′ 4.5″

(C) DIMENSIONS

Fig. 2-43. Two-bay stacked array.

Fig. 2-38 and is described in the section entitled *Mobile Antennas*. The length of the whip is 54 inches for 6 meters (52 MHz), 19.25 inches for 2 meters (146 MHz), 12.5 inches for 1 meter (222.5 MHz), and 6.5 inches for 0.70 meter (435 MHz).

Also popular for 1 and 2 meters is the 5/8-wavelength vertical whip with base-loading coil (see Fig. 2-35B and associated text for a description of the base-loaded whip). This antenna has a length of 48 inches for 2 meters (146 MHz) and 31.5 inches for 1 meter (222.5 MHz). Some hams have adapted commercial, base-loaded CB antennas for ⅝-wavelength use by revamping the coil in these antennas and telescoping the whip down.

EMERGENCY AND MAKESHIFT ANTENNAS

Circumstances—such as an emergency or a temporary shortage of supplies—occasionally dictate that some substitute for a standard antenna be used. In the past, hams have demonstrated exceptional ingenuity in pressing into service as antennas many unlikely devices. There is space here to discuss only a few such makeshifts. In a general sense, any metallic mass, which is insulated from ground, is mounted well in the clear, and can load the transmitter or pick up passing waves for the receiver, will serve as an emergency antenna. The efficiency of such a device depends upon many factors, such as physical size with respect to a quarter-wavelength, kind of metal, quality of insulation, effectiveness of coupling, and so on.

Zip-Cord Dipole

Figure 2-45A shows a simple dipole antenna made from ordinary two-parallel-conductor electric zip cord. The cord is simply split for a suitable length and the two halves pulled apart to form the half-wave top of the antenna, while the unsplit portion forms the parallel-wire feeder. The wire manufacturer intended this cord for 60 Hz AC use (lamp cords, etc.) and no information is easily had on its RF characteristics. It is reasonable to assume, however, that the plastic-insulated type will have lower losses than will the rubber-insulated type, and that the clear or white insulation is better than the colored type. For polyvinyl-chloride-insulated zip cord having two AWG #18 conductors, the impedance of the feeder portion of the antenna is approximately 44.6 ohms.

The conductors in zip cord are stranded wire. When using this cord, avoid breaking any of the strands. At the ends of each of the three legs of the zip-cord antenna, tin each strand separately, then solder them all together.

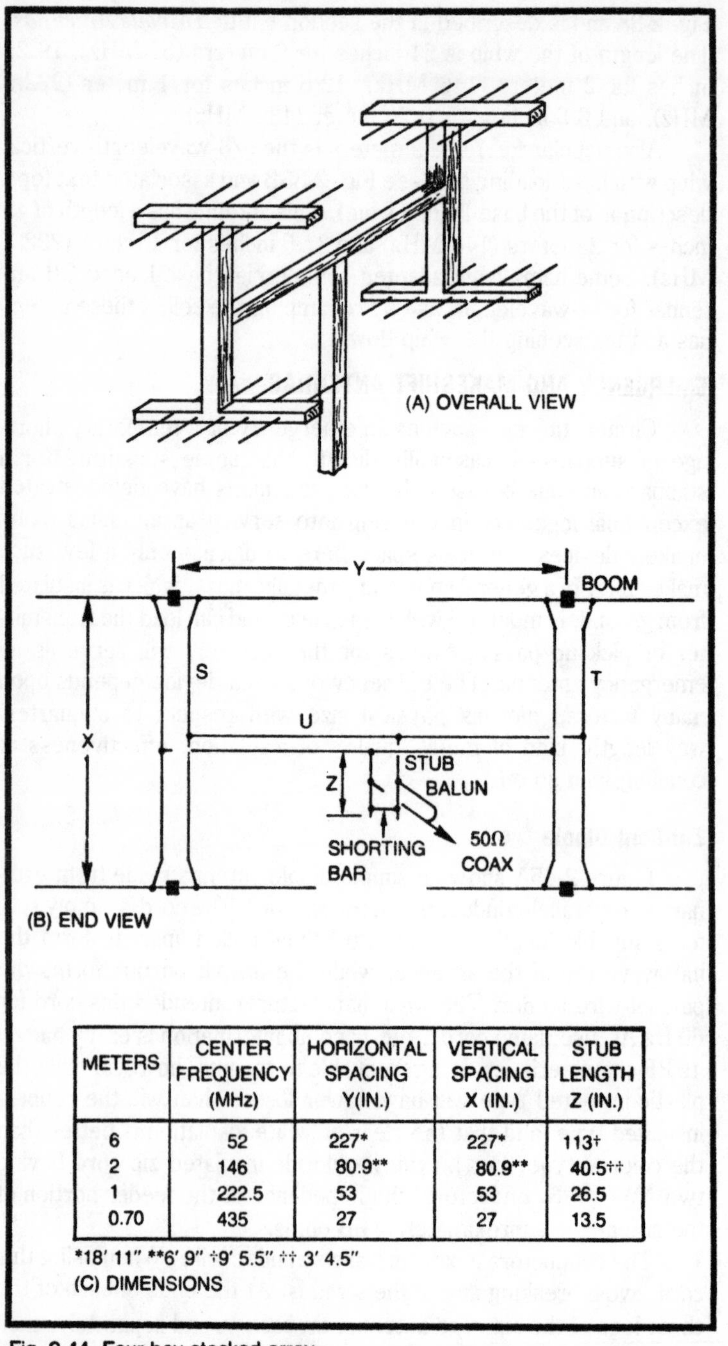

METERS	CENTER FREQUENCY (MHz)	HORIZONTAL SPACING Y(IN.)	VERTICAL SPACING X (IN.)	STUB LENGTH Z (IN.)
6	52	227*	227*	113†
2	146	80.9**	80.9**	40.5††
1	222.5	53	53	26.5
0.70	435	27	27	13.5

*18′ 11″ **6′ 9″ †9′ 5.5″ †† 3′ 4.5″

Fig. 2-44. Four-bay stacked array.

TV Rabbit Ears

Another simple, emergency dipole—this time of the adjustable V-type—is the familiar indoor rabbit-ear antenna sold for TV receivers (see Fig. 2-45B). This antenna is easily portable, its legs are telescoping, the V-angle is smoothly adjustable, and the antenna is easily rotated. While the rabbit-ear antenna is basically an indoor type, it can be placed outside in good weather. The simplest type that is most quickly adapted to emergency ham use is the type that has no auxiliary elements (such as transformers or coils), but comes with an attached 300-ohm ribbon.

A typical rabbit-ear antenna telescopes down to 10 inches per leg and out to 38 inches per leg. At 10 inches per leg, the antenna is a dipole at approximately 530 MHz; at 38 inches per leg, it is a dipole at approximately 150 MHz. (These figures assume an antenna leg diameter of 0.281 inch.)

TV or FM Antenna

To use an outdoor TV or FM receiving antenna as an emergency ham antenna, tie the two wires of the 300-ohm ribbon together at their lower ends, and operate the arrangement against ground as a Marconi antenna (see Fig. 2-45C). This setup is similar to the T-type Marconi antenna shown in Fig. 2-1C but without necessarily having the quarter-wave designation.

In a somewhat similar fashion, the two wires of an open-wire line that center feeds an antenna (see Fig. 2-45D) may be tied together at their lower ends to form a T-type Marconi antenna. This arrangement is often used for emergency operation of an antenna on a frequency lower than that for which it was designed (for example, operating an 80-meter antenna on 160 meters).

Miscellaneous Makeshifts

Other objects for which success as an antenna substitute has been reported include gutters and downspouts, metal roofs, lightning rods, window screens, wire fences, and metal window frames. The effectiveness of all such items as these depends upon their insulation, how high and how much in the clear they are situated, and how well energy can be coupled into them in transmitting or out of them in receiving. Because of the considerably lower amplitudes and the high sensitivity of receivers, most makeshift devices function surprisingly well as receiving antennas.

WHAT ABOUT RECEIVING ANTENNAS?

Most hams do not employ a separate receiving antenna. Instead, they switch the same antenna back and forth between trans-

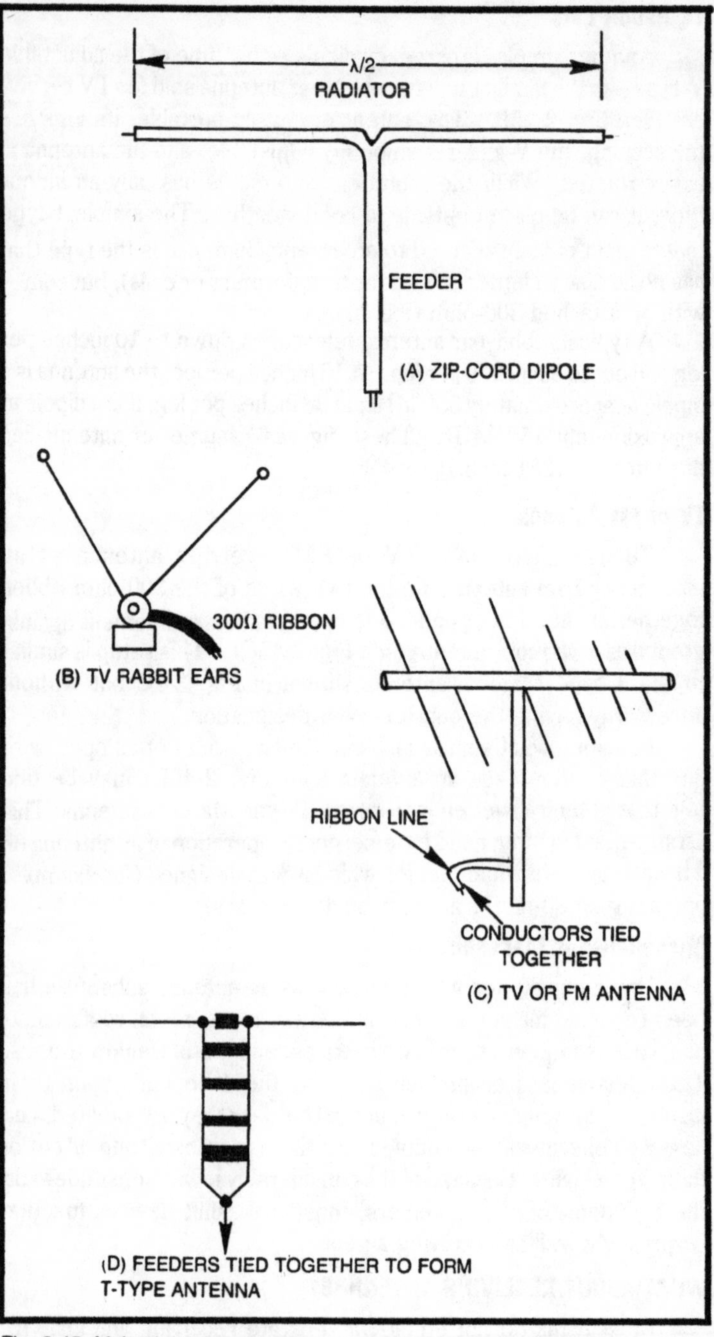

Fig. 2-45. Makeshift ham antennas.

mitter and receiver. This makes sense, for the antenna that transmits a signal efficiently to a certain point also receives efficiently from the same point. Most ham communication is point to point, not broadcast. The antenna is switched between transmitter and receiver by means of an SPDT or DPDT changeover switch or by means of a suitable relay operated either from the station's send/receive switch or from the receiver changeover switch (receiver off/transmitter on). In a transceiver, one or two sections of the transmit/receive switch (or relay) or the push-to-talk switch handle the antenna switching. Most modern receivers and transmitters have low-impedance antenna circuits, which means that a special coaxial RF-type relay is needed for antenna switching. A typical relay of this kind switches a 50-ohm coaxial line between transmitter and receiver, handles 1 kilowatt up to 400 MHz, is actuated either by 12V DC or 115V AC (depending upon model), and provides isolation of receiver during transmission of better than −100 dB.

When a separate receiving antenna is employed, it is usually simple—often a random horizontal length of wire or a long, vertical whip erected as high above ground as practicable and cut to a length which will afford acceptable performance in the bands favored by the user. Although the high sensitivity of the modern receiver insures that even a few microvolts of signal picked up by a simple antenna can result in satisfactory communications, the separate receiving antenna should not be slipshod. Moreover, it must be installed at a noise-free or, at least, low-noise point and preferably out of the inductive field of the transmitting antenna, to minimize noise interference and transmitter-receiver interaction. Also, since the receiving antenna unavoidably receives some strong signals from the transmitting antenna, some provision must be made to protect the receiver input circuit from burnout during transmitting intervals. The best remedy is to ground the receiving antenna during transmission; but, in lieu of a switch, this can sometimes be done automatically by connecting a suitable varistor (voltage-dependent resistor) between the receiving antenna and ground. The varistor looks like an open circuit to the weak signals picked up normally by the receiving antenna, but like a short circuit to the strong transmitter signal.

Chapter 3
Citizens Band Antennas

At the time of this writing, there are 20 million licensed citizens band (CB) stations in the United States; and by the time this book is in print, the number will have increased significantly. It is no wonder, then, that a number of manufacturers are supplying CB antennas—both base station and mobile—of every type and description. Since the price of these antennas is far from prohibitive, few CB operators will make their own antennas entirely from scratch except in an extreme emergency. Therefore, this chapter is devoted to a description of standard CB antennas and their characteristics, rather than to their design and construction. Its aim is to help the CB operator select the antenna that is best suited to his aims and environment.

Representative models have been selected for discussion, since naturally the multitude of CB antennas now being manufactured cannot be described individually in this space. For background information on omnidirectional and beam-type antennas, see Chapters 1 and 2.

THE CITIZENS BAND

The 440-kHz-wide class D citizens band consists of 40 channels extending from 26.965 MHz (channel 1) to 27.405 MHz (channel 40). Table 3-1 lists these channels and their corresponding frequencies. The center of this band lies at 27.185 MHz (channel 19). Like ham stations, CB stations may be either base (home) or mobile (vehicle).

Table 3-1. Class D CB Channels And Frequencies.

CHANNEL	FREQUENCY (MHz)	CHANNEL	FREQUENCY (MHz)
1	26.965	21	27.215
2	26.975	22	27.225
3	26.985	23	27.255*
4	27.005	24	27.235
5	27.015	25	27.245
6	27.025	26	27.265
7	27.035	27	27.275
8	27.055	28	27.285
9	27.065	29	27.295
10	27.075	30	27.305
11	27.085	31	27.315
12	27.105	32	27.325
13	27.115	33	27.335
14	27.125	34	27.345
15	27.135	35	27.355
16	27.155	36	27.365
17	27.165	37	27.375
18	27.175	38	27.385
19	27.185	39	27.395
20	27.205	40	27.405

*—In terms of frequency, channel 23 actually lies between channels 25 and 26 (thus: 25, 23, 26).

The Federal Communications Commission (FCC) has set a power-output maximum of 4 watts for all CB stations using amplitude modulation (AM) and 12 watts PEP (peak-envelope power) for those using single-sideband (SSB) modulation. This is *low* power, when compared with the maximum permitted other radio services (for example, the 4-watt CB rating is four-thousandths of the maximum power input allowed ham stations). The upshot of this is that the CBer has absolutely no RF energy to waste, so the CB antenna must radiate as efficiently as possible the energy delivered by the transmitter.

BASIC CB ANTENNA REQUIREMENTS

With CB power limited to the low value of 4 watts, effective communication depends largely upon antenna efficiency. That is, how strongly a station's signals are received at a given point and how great a distance can be covered depend entirely upon how well the antenna transfers RF energy from the transmitter to space. Consequently, the antenna is of more concern in CB work than in some other radio services. Important guideposts of CB antenna performance are as follows.

Mobile vs Base

CB antenna type, design, and selection often depend upon whether the antenna is to be used on a vehicle (mobile) or at a fixed location (base). The distinction turns out to be purely functional in some instances, however, since obviously a car-type antenna can be mounted on the roof of a building, and some antennas—such as the coaxial—may be either mobile or base. But, in general, base antennas tend to be larger, occasionally more complicated, and sometimes are of the beam type simply because there is more room for their installation at a base station. But whether the antenna is base or mobile, the CBer, forced to work with such a low power output, needs the best possible return for his antenna investment.

Angle of Radiation

The angle of radiation should be as low as practicable, in order to confine maximum energy to the ground wave. Since the sky wave has no value in local communications, energy in high-angle radiation is wasted. Common angles of radiation with good CB antennas vary from 15° to 45° from horizontal.

Directivity

Generally, a CB station communicates with other CB stations located anywhere in the service area. This *omnidirectional* transmission and reception requires an omnidirectional antenna. Such an antenna ideally has a circular radiation pattern, and at worst a pattern resembling a distorted circle or ellipse. It is the nature of a rod-type vertical antenna to exhibit such a pattern. A directional antenna radiates best in a straight line extending ahead from the front of the antenna and poorer, or ideally not at all, to the rear and off its sides. The directional, or *beam*, antenna therefore allows communication in only one direction, unless the antenna is rotated to aim it at a desired station in another direction.

A beam antenna provides gain over an omnidirectional antenna since the former concentrates the emitted RF energy largely in one direction. But because a beam antenna is more complicated than an omnidirectional one, such an antenna is more practical for a base station than for a mobile station. When point-to-point communication (that is communication between two points, rather than in all directions) occupies most of a station's time, a beam antenna is desirable, for it provides a link between the communication points and reduces interference with and from some stations not involved in the communications.

Polarization

For strongest signals, the transmitting and receiving antenna each should have the same polarization, either horizontal or vertical. When they are polarized differently (*cross polarization*), signal transfer is not nearly so good. In CB work, vertical polarization is common, because many of the stations in this service are mobile and a vertical antenna is the most practicable on a moving vehicle. Thus, it is usual to find base and mobile antennas both vertically polarized. Occasionally, a base-station antenna is designed for dual polarization when the station regularly must communicate with stations some of which use vertical polarization and some horizontal polarization. Rarely, such an antenna is switchable from one polarization to the other for the best signal at the receiving point.

Gain

An antenna is, of course, not an amplifier, so it cannot provide gain in the usually accepted meaning of the term. However, an antenna can be improved so that its output is a number of times higher than that of a simple reference antenna operated from the same transmitter, and this figure of improvement may be regarded as gain. Thus, a CB antenna that delivers twice the signal strength delivered by a basic antenna, such as a dipole, quarter-wave whip, or simple quarter-wave ground-plane antenna (most often the latter), exhibits a gain of 2 (or, expressed in another way, 3 dB).

There are only two ways of achieving antenna gain: (1) use of a beam antenna, and (2) increasing the low-angle energy in the radiation. The gain obtained by lowering the angle of radiation may be come about more simply, but does not equal the gain obtained with a beam. Gain ratings of CB antennas extend from 3 dB (i.e., two times) to 18 dB (i.e, 63.1 times), depending upon make and model.

The advantage of antenna gain should be obvious from this simple instance: It is illegal to increase CB transmitter power itself above the allowed 4 watts, but an antenna having 9 dB gain will effectively boost the power 7.94 times, that is, to 31.8 watts approximately—this is legal.

On a power basis,

$$\text{dB} = 10 \log_{10} (P_o/P_i) \qquad (3\text{-}1)$$

where P_i = input power, and
P_o = output power.

Table 3-2. Wavelengths In Inches For CB Channels.

DECIBELS VS. POWER RATIO

dB	POWER RATIO
1	1.26
2	1.58
3	1.99
4	2.51
5	3.16
6	3.98
7	5.01
8	6.31
9	7.94
10	10.00
11	12.59
12	15.85
13	18.95
14	25.12
15	31.62
16	39.81
17	50.12
18	63.09
19	79.43
20	100.00
21	125.90
22	158.49
23	199.53
24	251.19
25	316.23

Here, input power P_i is the power delivered to the antenna by the transmitter (maximum 4 watts for CB), and output power P_o is the effective power radiated by the antenna. For the reader's convenience, Table 3-2 gives power ratios corresponding to decibels between 1 and 25, and Table 3-3 gives decibels corresponding to power ratios between 1 and 500.

Additional discussion of antenna gain may be found in Chapter 1.

Bandwidth

While it is desirable to adjust the antenna at each frequency (as by varying the length of its elements), this is not often practicable. In the citizens band, one antenna usually must operate over the entire 40-channel spectrum with an acceptably low standing-wave ratio (SWR). This means that the antenna must exhibit reasonably broadband response.

Various design techniques serve to broadband an antenna. Some of thse involve some form of Q reduction. A high-Q antenna is

Table 3-3. Decibels VS Power Ratio.

POWER RATIO	dB
1	0
2	3.01
3	4.77
4	6.02
5	6.99
6	7.78
7	8.45
8	9.03
9	9.54
10	10.00
15	11.76
20	13.01
25	13.98
30	14.77
35	15.44
40	16.02
45	16.53
50	16.99
55	17.40
60	17.78
65	18.13
70	18.45
75	18.75
80	19.03
85	19.29
90	19.54
95	19.78
100	20.00
150	21.76
200	23.01
250	23.98
300	24.77
350	25.44
400	26.02
450	26.53
500	26.99

a sharply tuned (narrow-band) antenna, whereas a low-Q antenna tunes broadly, covering more frequencies but with somewhat reduced efficiency. For compromise broadband operation, the CB antenna is simply designed for the center of the band (channel 19: 27.185 MHz) and its resulting performance at the lower and higher frequencies accepted.

It is common for antenna bandwidth to be expressed in terms of frequency points above and below resonance, at which the SWR just reaches 2:1 (for example, 700 kHz below resonance and 700 kHz above resonance). Fig. 3-1 shows an example in which a center-band antenna exhibits SWR of 1.1:1 at resonance and 2:1 at 26.485 MHz and 27.885 MHz (700 kHz below and above resonance).

Physical Dimensions

Preferred lengths for CB antenna elements are full-wave, five-eighths-wave, half-wave, and quarter-wave. Naturally, a full-wave vertical antenna (36 feet 6 inches for channel 1) could be accommodated at only a few base stations and on no vehicles that might pass under obstructions. Therefore, the shorter types are the most practical. The required length of an antenna element may be calculated in terms of the desired frequency, thus:

$$l = 11{,}811/f \qquad (3\text{-}2)$$

where l = required length in inches, and
f = desired frequency in megahertz.

This formula does not take into consideration the ratio of element diameter to half-wavelength; however, it is sufficiently accurate for practical purposes. For convenience, Table 3-4, based on this formula, gives lengths in inches for common sizes of antenna elements.

When the antenna element must be physically shorter than a quarter-wavelength, as when vertical space is restricted, a loading coil can be inserted into it to increase the electrical wavelength to a quarter-wave. The coil is inserted at either the bottom of the antenna (*base loading*), center of antenna (*center loading*), or top of

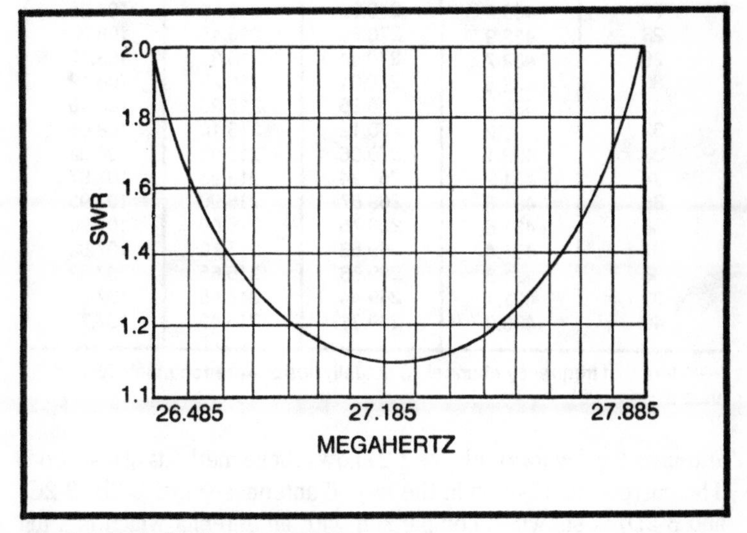

Fig. 3-1. Typical antenna response.

Table 3-4. Power Ratio VS Decibels.

WAVELENGTHS IN INCHES FOR CB CHANNELS

CHANNEL	λ	5λ/8	λ/2	λ/4
1	438	273.7	219	109.5
2	437.8	273.6	218.9	109.45
3	437.7	273.5	218.8	109.42
4	437.4	273.4	218.7	109.35
5	437.2	273.25	218.6	109.30
6	437.0	273.12	218.5	109.25
7	436.9	273.06	218.45	109.22
8	436.5	272.81	218.25	109.12
9	436.4	272.75	218.20	109.10
10	436.2	272.62	218.10	109.05
11	436.1	272.56	218.05	109.02
12	435.7	272.31	217.85	108.92
13	435.6	272.25	217.80	108.90
14	435.4	272.12	217.70	108.85
15	435.3	272.06	217.65	108.82
16	434.9	271.81	217.45	108.72
17	434.8	271.75	217.40	108.70
18	434.6	271.62	217.30	108.65
19	434.5	271.56	217.25	108.62
20	434.1	271.31	217.05	108.52
21	433.9	271.18	216.95	108.47
22	433.8	271.12	216.90	108.45
23*	433.3	270.81	216.65	108.32
24	433.7	270.06	216.85	108.42
25	433.5	270.93	216.75	108.37
26	433.2	270.75	216.60	108.30
27	433.0	270.62	216.50	108.25
28	432.9	270.60	216.45	108.22
29	432.7	270.44	216.35	108.17
30	432.5	270.31	216.25	108.12
31	432.4	270.25	216.20	108.10
32	432.2	270.12	216.10	108.05
33	432.1	270.06	216.05	108.02
34	431.9	269.94	215.95	107.97
35	431.8	269.87	215.90	107.95
36	431.6	269.75	215.80	107.90
37	431.4	269.62	215.70	107.85
38	431.3	269.56	215.65	107.82
39	431.1	269.44	215.55	107.77
40	430.9	269.31	215.45	107.72

*—In terms of frequency, channel 23 actually lies between channels 25 and 26 (thus :25, 23, 26).

antennas (*top loading*). Figure 3-2 shows these methods of insertion. The current distribution in the loaded antennas (Figs. 3-2B, 3-2C, and 3-2D) is shown in comparison with an antenna which is a full quarter-wave long (Fig. 3-2A). Base loading is the most solid arrangement, mechanically, but top loading introduces the least distortion of the quarter-wave current distribution and of the radiated field. However, top loading is the least desirable arrangement, mechanically, since it gives rise to much swinging of the antenna when the vehicle is moving (which can result in signal flutter) and exposes the coil to damage from blows by obstructions. The metal element (whip) of the antenna may be of the telescoping type, to permit close adjustment of length.

Elements for the class D CB antenna are usually made of hollow stainless steel or aluminum tubing. They are also made of fiberglass tubing or rod in which a conductor is embedded or around which a coil is wound. The elements may be of the telescoping type, to permit close adjustment of length. For the same fraction of a wavelength, a fiberglass antenna is shorter than a metal one. The reason for this is that the fiberglass—on which the conductor rests and which, in the form of a protective coat, surrounds the conductor—changes the velocity factor of the conductor. Thus, a quarter-wave fiberglass antenna is only 96 inches long, whereas a quarter-wave stainless steel antenna is 108 inches long.

Metal elements of a CB antenna sometimes are coated with a thin coat of plastic protective material—by the user—to guard against corrosion when the antenna is in a marine environment. But when this is done, the user must remember that the plastic alters the velocity factor and accordingly the wavelength of the coated element.

Impedance and Transmission Line

Many omnidirectional CB antennas operate directly from a 50- or 52-ohm coaxial transmission line. This is a convenience, as most

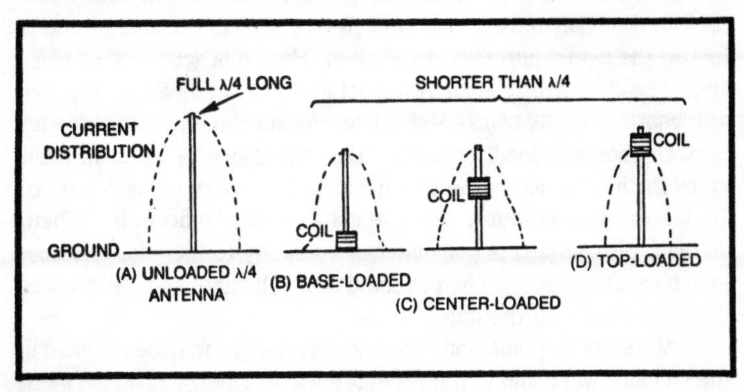

Fig. 3-2. Loaded antennas.

CB transceivers have 50-ohm output, so no matching device is needed.

The impedance of the radiator (driven element) of beam antennas is somewhat less than 50 ohms and requires some form of matching device between the radiator and the commonly used coaxial cable. Widely used devices are the ***beta match*** and ***gamma match***. The beta match (Fig. 3-3A) consists of a small, adjustable matching stub composed of two short rods or lengths of tubing (*X* and *Y*) and a sliding, short-circuiting strip (*Z*). The rods are fastened to the two halves of the dipole radiator at its center, as is also the coaxial cable. The strip slides along the grounded metal boom of the array and simultaneously along the two rods. The strip thus is center grounded to the boom for DC (such as a lightning stroke), but it is effectively an open circuit at RF. In adjustment, the strip is moved to the point which affords lowest SWR. The gamma match (Fig. 3-3B) is mounted on one side of center of the radiator and consists of a sliding rod or tube (*S*) held by a metal clamp (*T*), and a small variable capacitor (*C*). Some versions of the gamma match omit the capacitor. During adjustment, the rod is slid trombone fashion through the clamp to vary its effective length (and the capacitor is tuned at the same time) for lowest SWR. When the capacitor is omitted, the sliding rod is connected directly to the outer conductor of the coaxial cable.

Standing-Wave Ratio

Under ideal conditions, all of the RF output of a CB transmitter travels down the transmission line—as the ***incident wave***—and is radiated by the antenna. But when there is not a perfect impedance match between antenna and line, a portion of the energy is returned down the line—as the *reflected wave*—from the antenna connection back to the transmitter. This reflected component is energy that is lost to the useful purpose of radiation. How this action takes place should be clear from the following details: The antenna acts as a load at the far end of the transmission line. For maximum power transfer between line and load (antenna), the impedance of the load must equal the impedance of the transmission line and must be resistive. Under these conditions, an antenna functions efficiently. When, instead, the antenna is not matched correctly to the line, this mismatch causes energy to be reflected down the line, and this creates standing waves on the line.

At a suitable point along the line, the energy may be sampled in the incident wave and in the reflected wave, and the ratio between

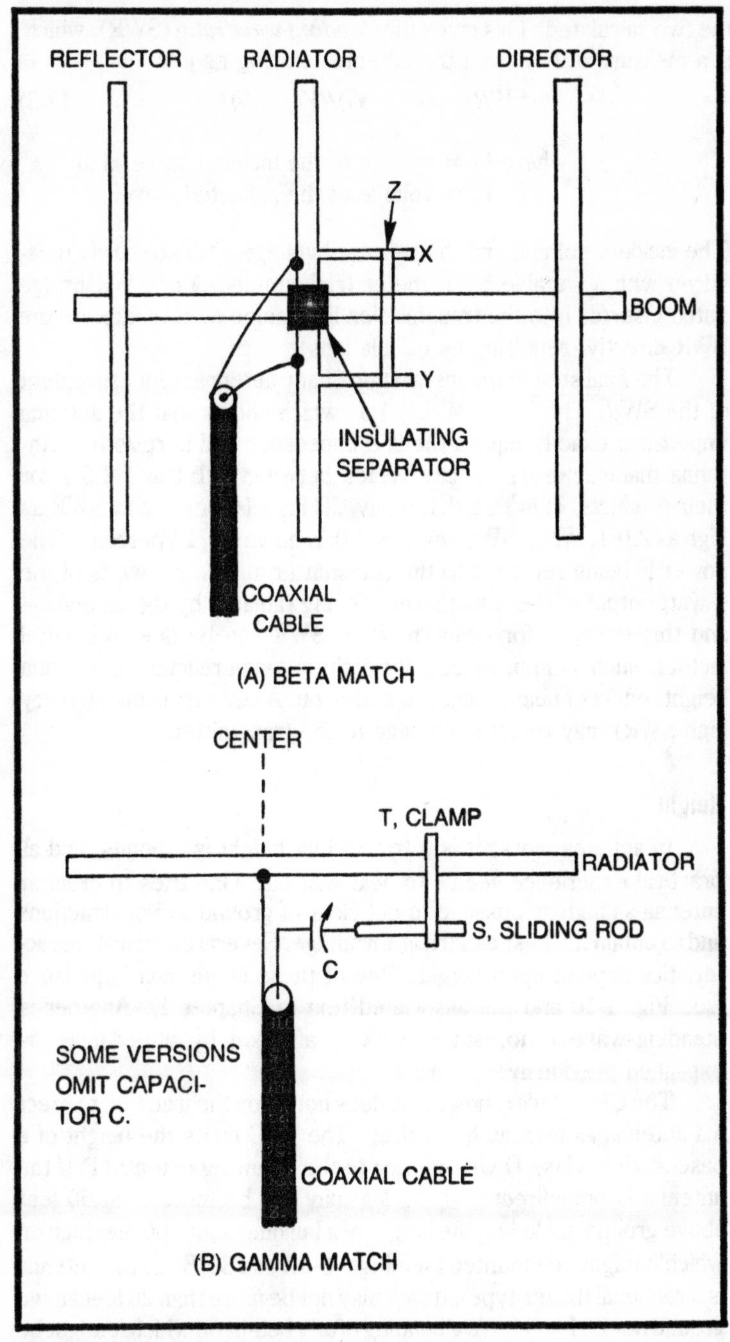

Fig. 3-3. Matching devices.

the two calculated. This gives the *standing-wave ratio* (SWR), which is a measure of how well the antenna is doing its job:

$$SWR = (V_i + V_r)/(V_i - V_r) \quad (3\text{-}3)$$

where V_i = voltage of the incident wave, and
V_r = voltage of the reflected wave.

The incident voltage and the reflected voltage both are easily measured with a suitable SWR meter (reflectometer) or SWR bridge patch-inserted into the transmission line (some meters indicate the SWR directly, requiring no calculations).

The final step in the installation of any antenna is measurement of the SWR. The ideal SWR is 1:1, which shows that the antenna impedance exactly equals the line impedance and is resistive. Antenna manufacturers specify values between 1.1:1 and 2.5:1 for their products. It is said that many CB installations run at SWR as high as 2.0:1. When SWR reaches 3.0:1, however, 25 percent of the power is being returned to the transmitter and only 3 watts of the 4-watt output of the transmitter is being radiated by the antenna—and this is cause for concern. Poor SWR can be due to several factors, such as impedance mismatch, antenna reactance, antenna height, effect of nearby objects, and so on. A severe mismatch (very high SWR) may result in damage to the transmitter.

Height

In antenna work, it is a truism that height is a bonus, and all practical experience seems to bear that out. One tries to erect an antenna as high as possible to get clear of ground and obstructions and to obtain the best electrical advantage. Several electrical characteristics depend upon height. One of these is antenna impedance (see Fig. 1-10 and the associated text in Chapter 1). Another is standing-wave ratio, since SWR is affected by impedance, as explained previously.

The CB installer, however, does not enjoy the freedom to erect his antenna as high as he desires. The FCC limits the height of a base-station, class D CB antenna to the following extent: (1) If the antenna is omnidirectional, its *top* may not be more than 60 feet above ground, including the height of a building under 60 feet high on which it might be mounted (see Figs. 3-4A and 3-4B). If the antenna is directional (beam-type), its *top* may not be more than 20 feet above ground nor 20 feet above a building of *any* height on which it might be mounted (see Figs. 3-4C, 3-4D, and 3-5B). If the antenna is

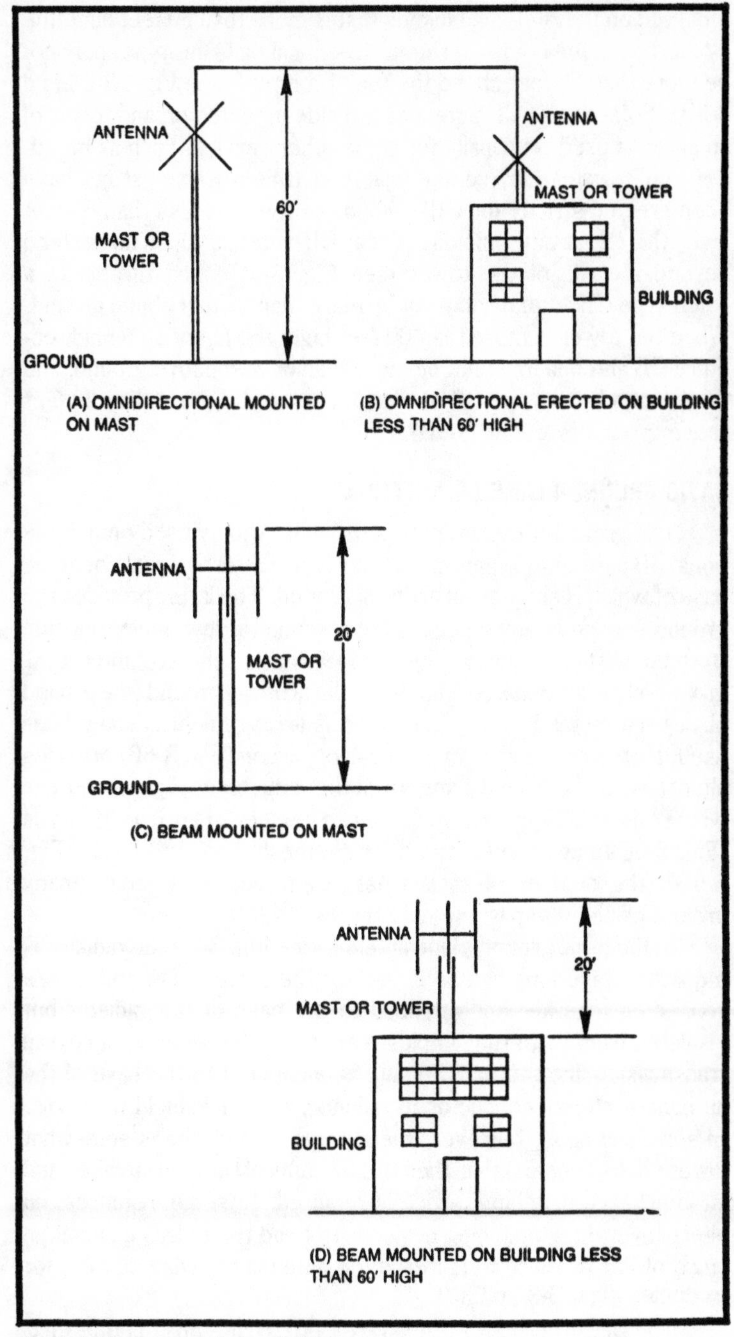

Fig. 3-4. Maximum antenna heights.

mounted on the roof of a building that is more than 60 feet high, the *top* of the antenna (whether omnidirectional or beam-type) may not be more than 20 feet above the top of the roof (see Figs. 3-5A and 3-5B). Sometimes, CB antennas are side mounted on and tower of an antenna used principally for some other service, such as broadcast, police, amateur, commercial, etc. (this tower must not have been erected strictly for CB). When the tower is less than 60 feet high, the *top* of any omnidirectional CB antenna must not extend beyond the top of the tower (see Fig. 3-5C), and the *top* of a beam-type CB antenna may not be more than 20 feet above ground. When the tower is more than 60 feet high, the *top* of an omnidirectional CB antenna must not be more than 60 feet above ground, and the *top* of a beam-type CB antenna must not be more than 20 feet above ground (see Fig. 3-5D).

BASIC GROUND-PLANE CB ANTENNA

The ground-plane antenna is the most widely used omnidirectional CB base-station antenna. It consists of a vertical radiator at the base of which is situated an artificial ground. The latter provides the ground needed at any height of the antenna for low-angle radiation from the vertical radiator. Figure 3-6 illustrates the evolution of the ground-plane antenna. At Fig. 3-6A, the artificial ground (the ground plane) is a metal disc or plate, and this arrangement is sometimes used. However, a disc of metal mesh or screen (Fig. 3-6B) provides almost as good a ground plane and at the same time is lighter and less wind resistant. Going a step further, a number of horizontal radials (Fig. 3-6C) may be substituted for the mesh disc. And finally (Fig. 3-6D), the number of radials has been reduced to four (many ground-plane antennas use only three).

In the basic ground-plane antenna (see Fig. 3-7), the radiator is a quarter-wave long and so is each of the radials. The radials are connected together and attached to the base of the radiator but insulated from it. In the simplest version of this antenna, a coaxial transmission line (50 or 52 ohms) is connected to the base of the antenna—internal conductor to radiator, external shield to radials. In some versions, however, the antenna impedance is somewhat lower (25 to 35 ohms) than the 50 to 52 ohms of the coaxial cable, and an impedance-matching device is required. In some instances, an alternative to the matching device is to bend the radials down at an angle of 35° or more to increase the antenna impedance. See, for example, Figs. 3-8 and 3-9.

The performance of other types of CB antennas is often expressed in terms of the performance of this basic ground-plane

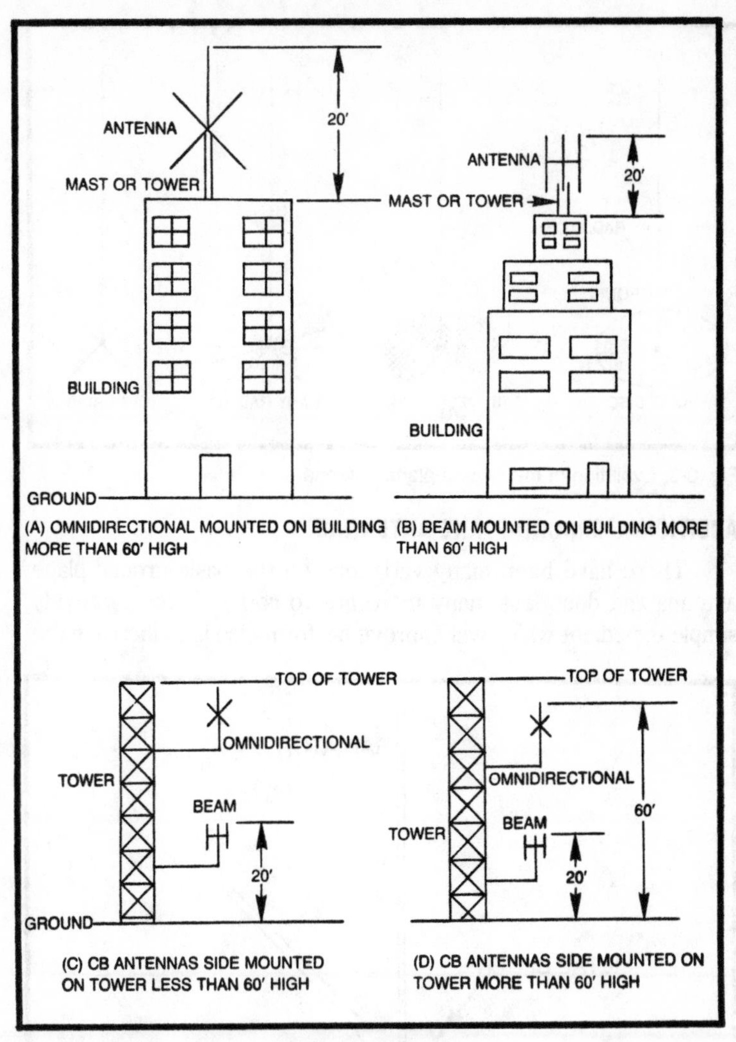

Fig. 3-5. Maximum antenna heights (continued).

antenna. When this is done, the gain of the basic ground-plane antenna is taken as zero decibel (0 dB). If a given antenna is said to have a gain of 3 dB, for example, the effective radiated power of that antenna is approximately 2 times that of the basic ground-plane antenna (which means that the 4 watts output of a CB transmitter is made to look like 8 watts). See Tables 3-2 and 3-3 for comparisons between decibels and power ratios.

For further fundamental discussion of the ground-plane antenna, see Chapter 2.

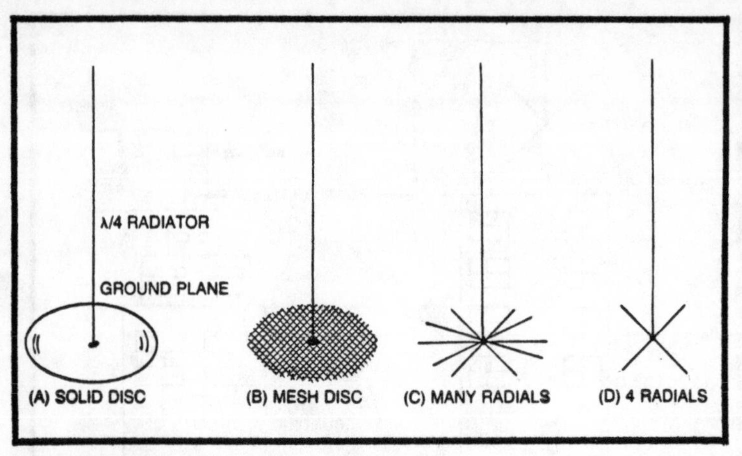

Fig. 3-6. Evolution of the ground-plane antenna.

ADDITIONAL GROUND-PLANE ANTENNAS

There have been many variations on the basic ground-plane antenna and doubtless many more are to come. A comparatively simple expedient which will improve performance is to increase the

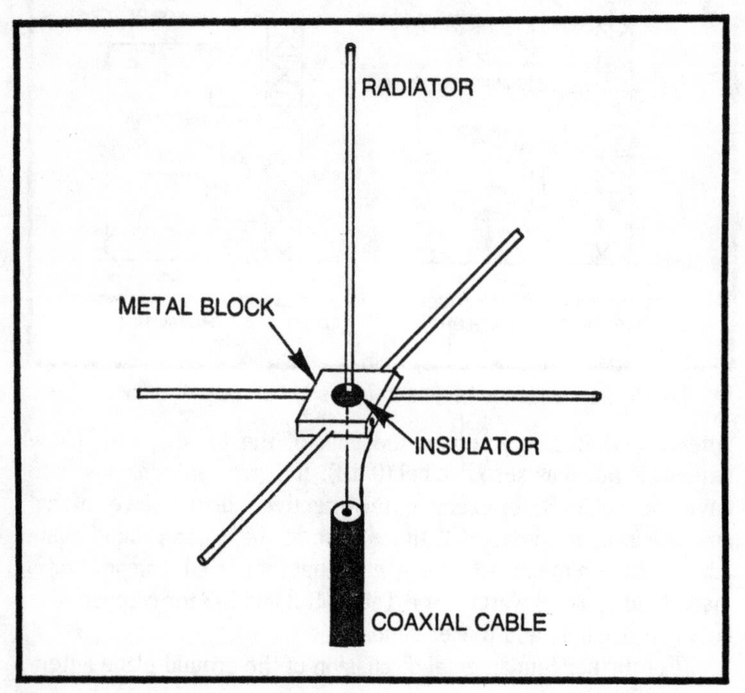

Fig. 3-7. Basic ground-plane antenna.

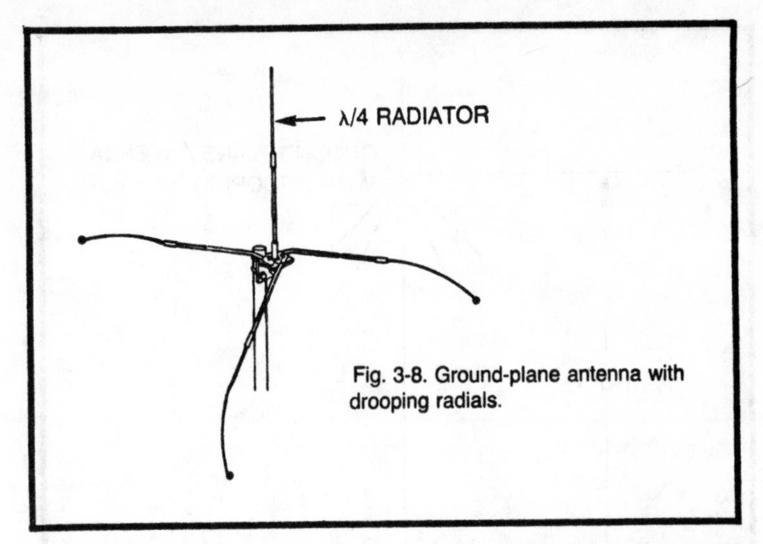

Fig. 3-8. Ground-plane antenna with drooping radials.

length of the radiator: A half-wave radiator will provide a gain of 1.76 dB or better, and a 5/8-wave radiator will provide a gain of 3 to 5 dB or better.

Drooping Radials

The drooping-radial version of the ground-plane antenna not only provides an impedance match between antenna and coaxial transmission line (the match being proportional to the droop angle), but also lowers the angle of radiation, thereby providing effective gain. Improved performance therefore may be expected with the slightly drooped model shown in Fig. 3-8, and will greater improve with the one shown in Fig. 3-9. The antenna shown in Fig. 3-8

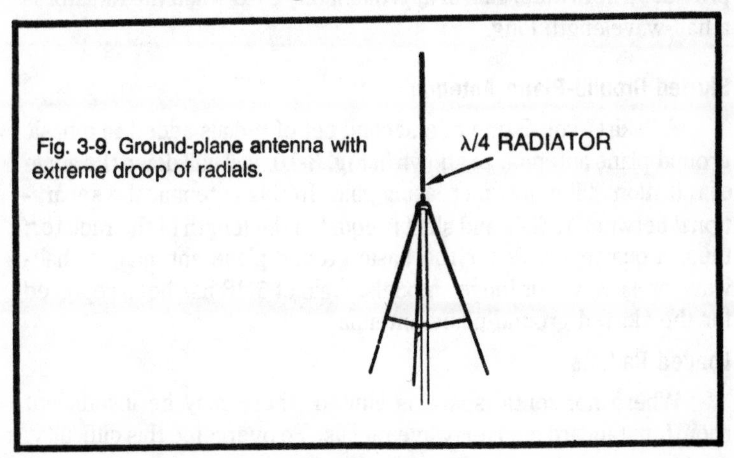

Fig. 3-9. Ground-plane antenna with extreme droop of radials.

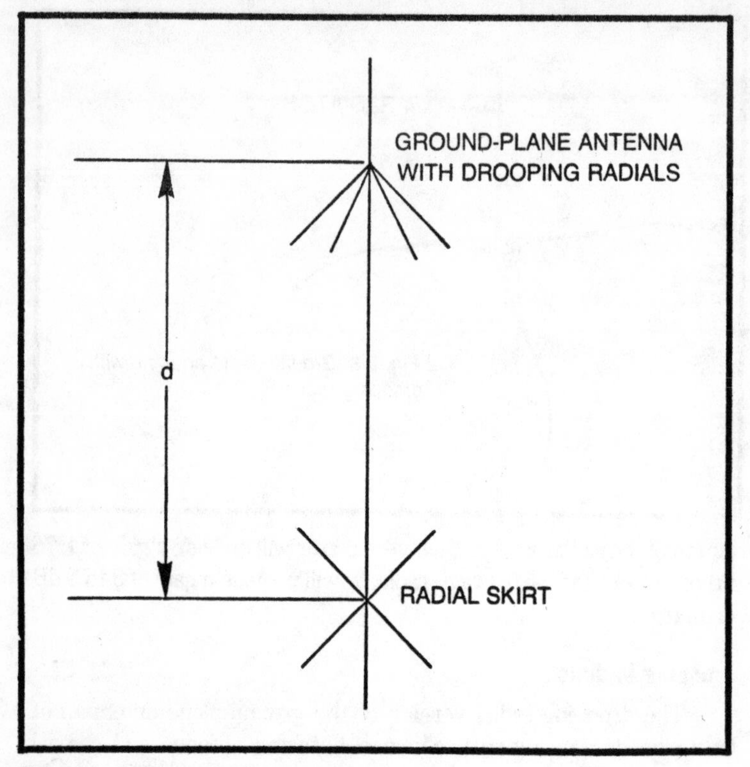

Fig. 3-10. Skirted ground-plane antenna.

permits a close match to 50-ohm coax, with no other matching medium (such as a transformer or stub). The one shown in Fig. 3-9 provides a horizontal gain of approximately 5 dB when the radiator is a half-wavelength long.

Skirted Ground-Plane Antenna

A "skirt" consisting of a second set of radials added to a basic ground-plane antenna, as shown in Fig. 3-10, will pull down the angle of radiation still more, increasing gain. In this antenna, the separation *d* between radials and skirt is equal to the length of the radiator; thus, a quarter-wave for the basic ground-plane antenna, or half-wave or ⅝-wave for longer models. Gain of 5 dB has been reported for the skirted ground-plane antenna.

Loaded Radials

Where horizontal space is limited, there may be insufficient room for standard quarter-wave radials. To overcome this difficulty, ground-plane antennas are available with short radials. To make the latter work, a loading coil is inserted into each radial (see Fig. 3-11) to increase the electrical length of the radial to a quarter-wave. Occasionally, a shortened radiator (where allowable height is restricted) also contains a loading coil. The radiator whip of a loaded car-top antenna or mirror-mounted antenna can be as short as 18 inches.

Capacitance Loading

A capacitance hat on the top of an antenna lowers the angle of radiation, for increased range, and at the same time reduces static noise in reception. A common hat on a ground-plane antenna consists of three short radials at the top of the radiator (see Fig. 3-12A). Other hats are the cloverleaf type (Fig. 3-12B), cloverleaf-ball type (Fig. 3-12C), diamond-loop type (Fig. 3-12D), and hexagonal-loop (hex-loop) type (Fig. 3-12E). For additional information regarding capacitance loading, see Fig. 2-36 and the associated text in Chapter 2.

Dual Antennas for Directivity

Two stationary ground-plane antennas can be operated together as a directional array, as shown in Fig. 3-13. The two

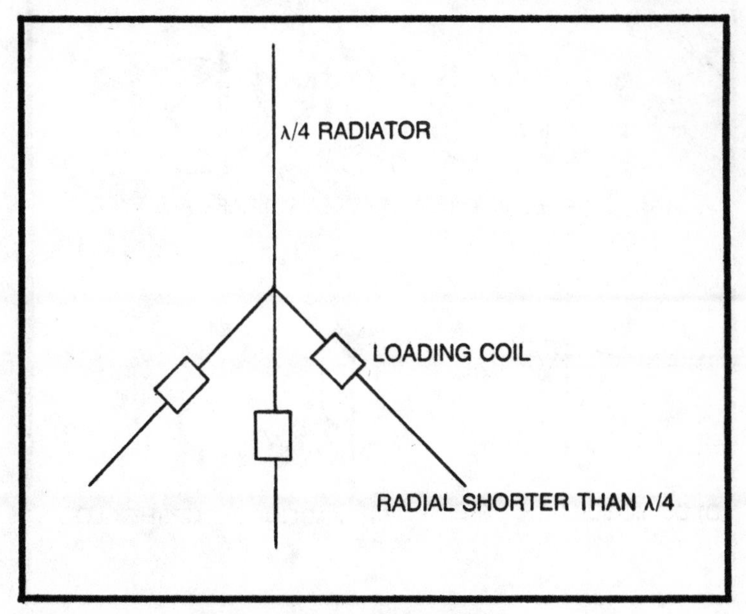

Fig. 3-11. Ground-plane antenna with loaded radials.

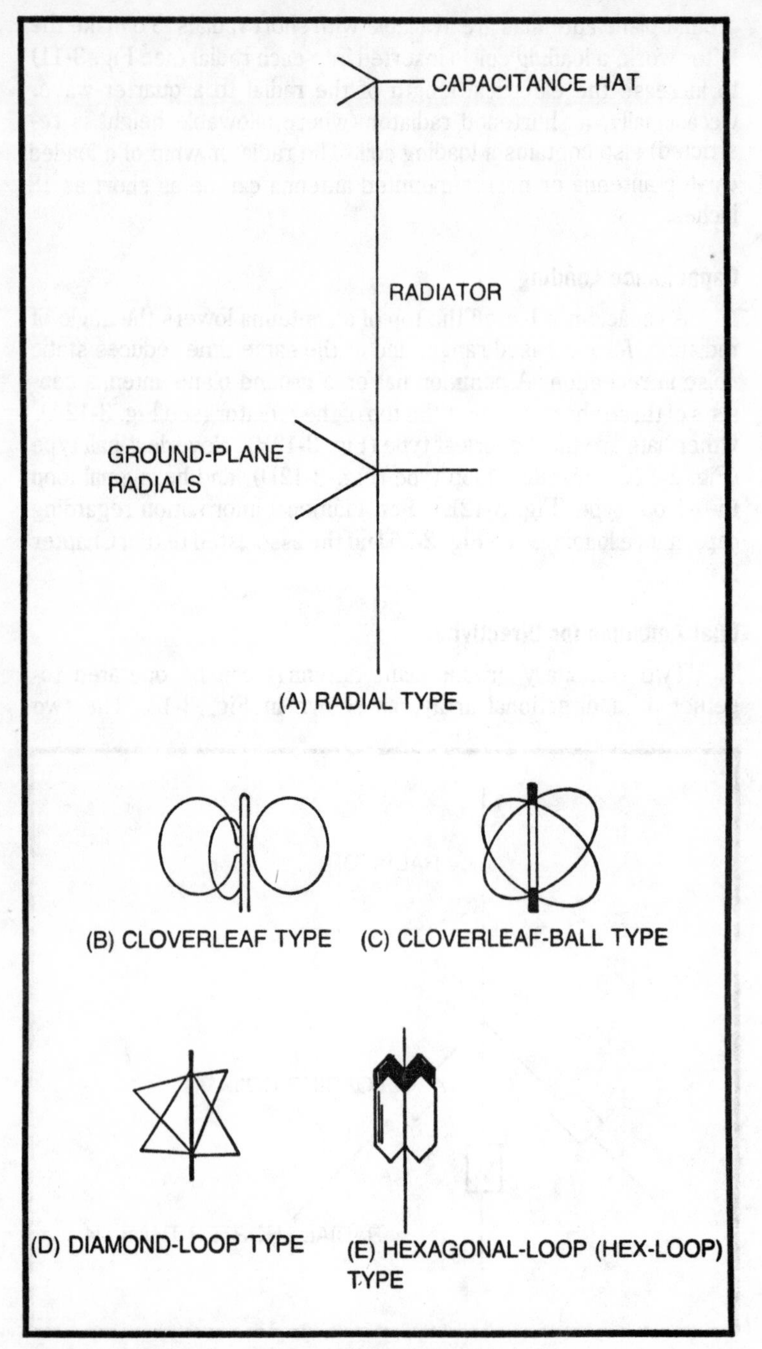

Fig. 3-12. Ground-plane antenna with capacitance hat.

antennas are fed simultaneously through a coaxial-type phasing harness, and the directivity is determined by the phase of the signals fed to the antennas through this harness.

In Fig. 3-13A, the antennas are separated by a half-wavelength, and equal-length sections of the harness feed the antennas in phase. The resulting pattern is a relatively narrow figure-8 broadside to the line of the antennas, and the gain is approximately 3.86 dB. When a half-wavelength is added to the rear section of the harness (Fig. 3-13B) and the antennas are a half-wavelength apart, the resulting pattern is end fire, and the gain is approximately 2.3 dB. When the antennas are a quarter-wavelength apart and a quarter-wavelength is added to the front section of the harness (Fig. 3-13C), the resulting pattern is cardioid through the front antenna. Where there is room to mount two antennas, this arrangement offers some of the versatility of a rotary beam without demanding the mechanical intricacies of the rotary antenna.

BASE-STATION VERTICAL

A tall vertical antenna affords a lower angle of radiation than that provided by a quarter-wave vertical, and accordingly provides gain. For example, a half-wave vertical gives an angle of approximately 30° (against approximately 50° for the quarter-wave ground-plane antenna) and gain of approximately 1.8 dB over the ground-plane; and a ⅝-wave vertical gives an angle of approximately 15° and gain of 3 to 4 dB over the ground-plane. Beyond ⅝-wavelength, the angle of maximum radiation again rises, and gain falls.

The omnidirectional radiation pattern of the tall vertical—together with the gain of this antenna, almost zero horizontal space requirement, and the fact that the vertical requires no ground-plane radials (some manufacturers include *short* radials to decouple the coaxial cable)—makes this antenna attractive to many CBers. Its increased height, however, restricts its use to base stations, though it can be telescoped and carried to a remote location for temporary field use. A ⅝-wave vertical for channel 1, the lowest CB frequency, is approximately 23 feet high. (See Table 3-4 for heights in inches of antennas.) A typical ⅝-wave vertical antenna is shown in Fig. 3-14A.

Several methods are available for matching a 50-ohm coaxial line to a ⅝-wave vertical antenna: In Fig. 3-14B, an autotransformer formed by a single-turn horizontal coil (or loop) is attached to the base of the radiator and shunted with a small variable capacitor, VC. The resulting *LC* circuit is tuned to resonance by adjusting the capacitor. The center conductor of the coax is connected to a sliding

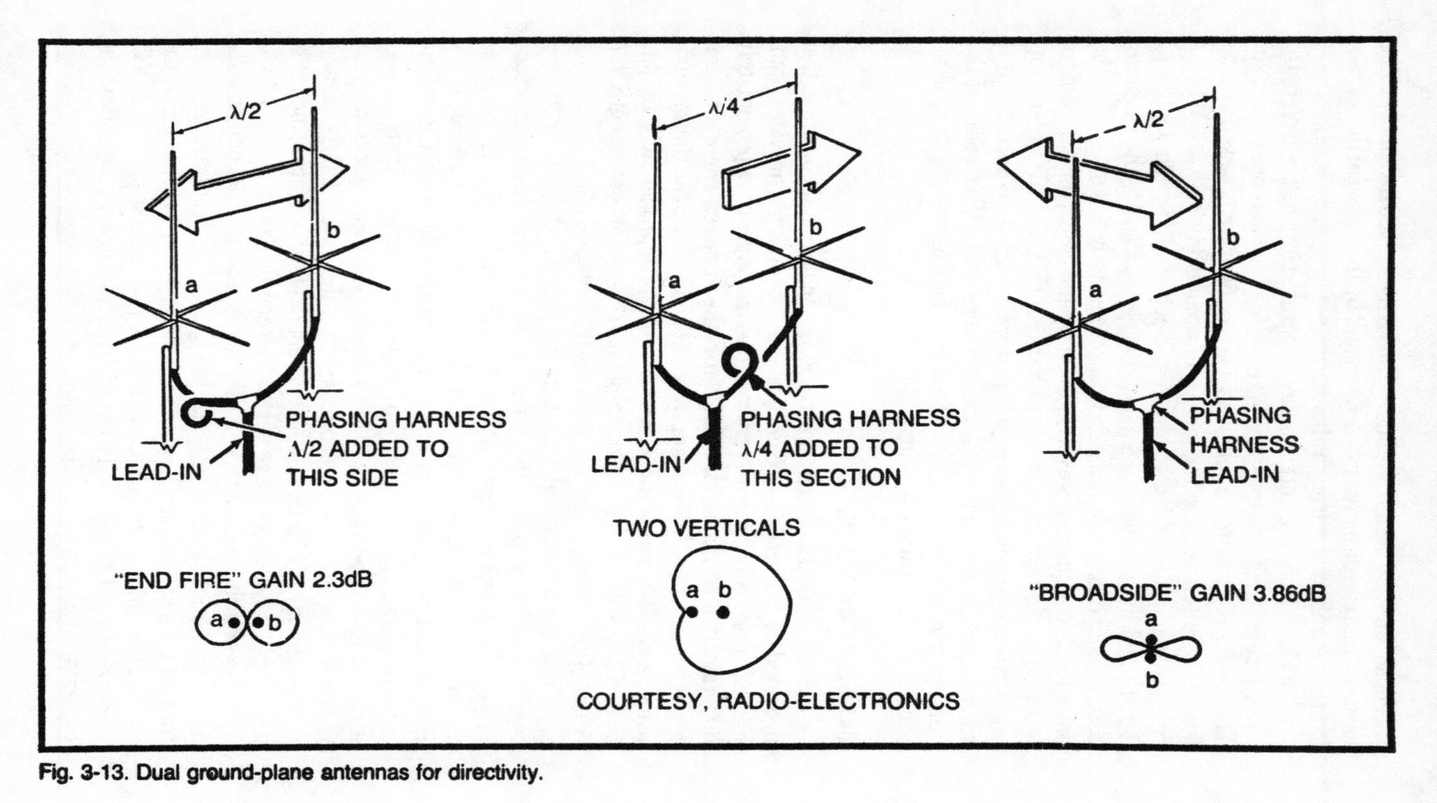

Fig. 3-13. Dual ground-plane antennas for directivity.

contact of the coil and may be positioned for the 50-ohm point on the coil. In commercial antennas of this type, the variable capacitor is a weatherproof concentric-cylinder-type unit. In some arrangements, the capacitor is omitted and the coil alone employed simply as an autotransformer without tuning. In Fig. 3-14C, a gamma match bent into a circle is used to match the impedances. Here again, a concentric capacitor is employed. A comparison of Fig. 3-14C with Fig. 3-3B will show how the standard, straight-line-type gamma match (Fig. 3-3B) has been converted into the circular type. Obviously, the antenna cannot conveniently be adjusted separately for each channel frequency, so either the loop tuning or the gamma match must be set for the most-used frequency; or, if all channels are used, set it for the center of the band (27.185 MHz, channel 19).

COAXIAL ANTENNA

Another excellent omnidirectional, low-angle antenna is the *coaxial* antenna, with a gain of approximately 1.8 dB. Since the total length of this antenna is a half-wavelength, the coaxial type is best suited—in 27 MHz service—to base-station use. Like the straight vertical antenna, the coaxial antenna requires almost zero horizontal space and is low profile. Also like the vertical, it needs no ground plane (although some manufacturers supply it with a ground plane for further lowering of the angle of radiation), nor does it need a ground connection.

The coaxial antenna is a vertical adaptation of the simple dipole. Its construction is shown in Fig. 3-15A. In this antenna, the center conductor of a coaxial cable is, in effect, extended a quarter-wavelength beyond the end of the cable, and the outer shield of the cable is connected to a quarter-wavelength metal sleeve that surrounds the cable a short distance. This antenna accordingly is a half-wave, center-fed vertical in which a quarter-wave section is the inner-conductor extension (the radiator in Fig. 3-15A), and the other quarter-wave section is the metal sleeve. The cable impedance may be 50 to 53 ohms. Incidentally, the sleeve shields the cable and forestalls the induction of current in the outer conductor of the cable by energy radiated from the antenna.

This arrangement provides a compact vertical antenna which gives low-angle radiation. In manufactured models, the inner-conductor extension (radiator) is a rod or whip, often telescoping for close adjustment of length, and is insulated from the sleeve through the center of which the cable passes. The sleeve may be either aluminum, copper, brass, or stainless steel. The coaxial antenna is

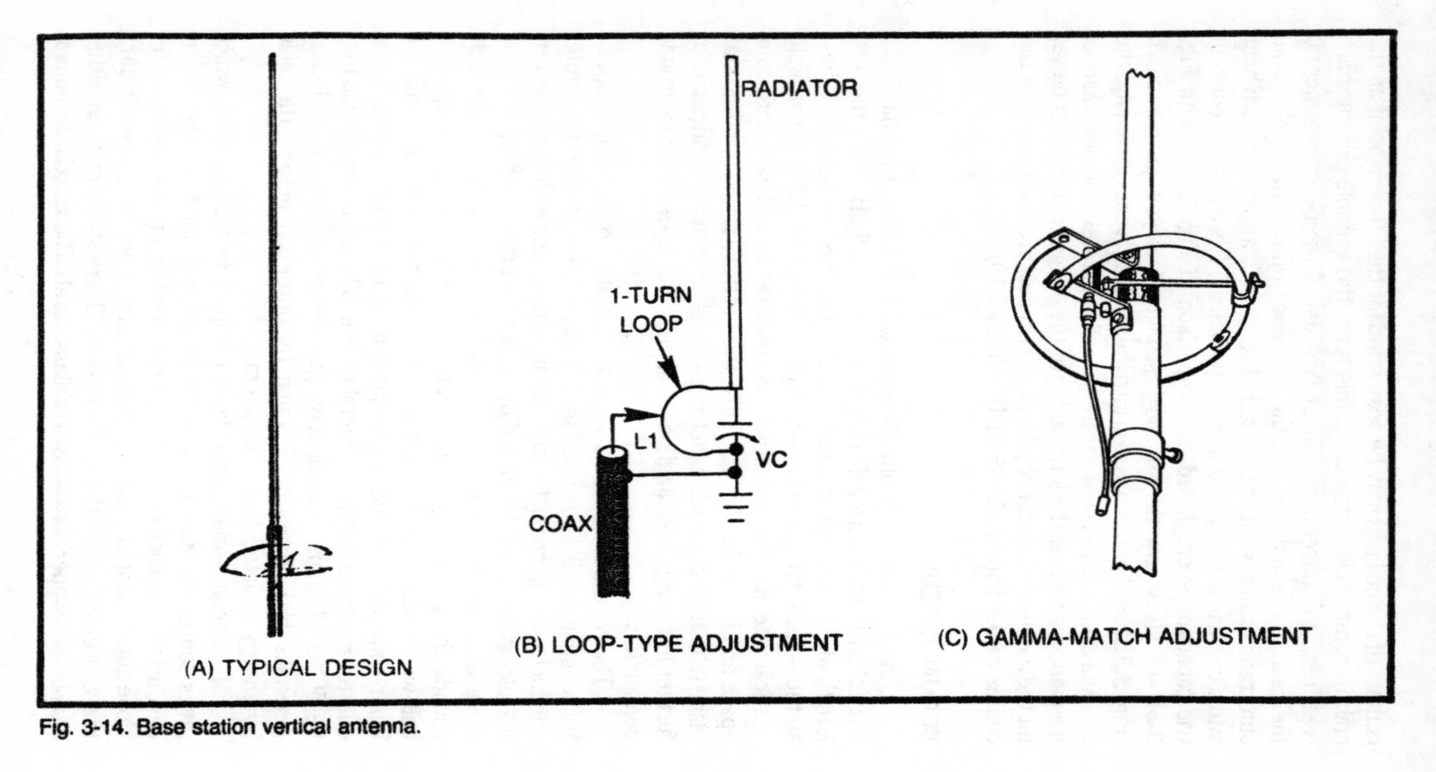

Fig. 3-14. Base station vertical antenna.

erected usually by fastening the lower end of the sleeve to the top of a pole or mast by means of an insulating bushing having a clearance hole for the cable. This mounting preserves the slender profile which is characteristic of well-built vertical antennas.

Figure 3-15B gives the radiator and sleeve lengths (in both inches and feet) for the top channel (40), center channel (19), and bottom channel (1). In the last column, the total length is given.

In commercial coaxial antennas, the radiator sometimes is shortened physically and provided with a center loading coil.

BASE-STATION BEAM ANTENNAS

Of all types of antennas, the *beam* antenna provides the highest gain and the best discrimination against unwanted signals—but in

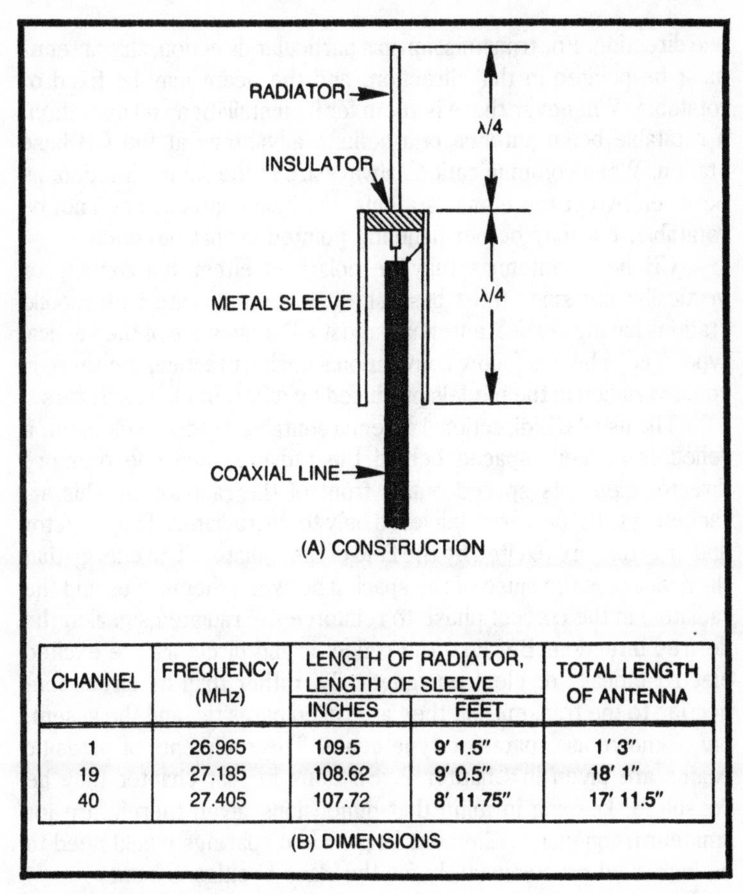

(A) CONSTRUCTION

CHANNEL	FREQUENCY (MHz)	LENGTH OF RADIATOR, LENGTH OF SLEEVE		TOTAL LENGTH OF ANTENNA
		INCHES	FEET	
1	26.965	109.5	9′ 1.5″	1′ 3″
19	27.185	108.62	9′ 0.5″	18′ 1″
40	27.405	107.72	8′ 11.75″	17′ 11.5″

(B) DIMENSIONS

Fig. 3-15. Coaxial antenna.

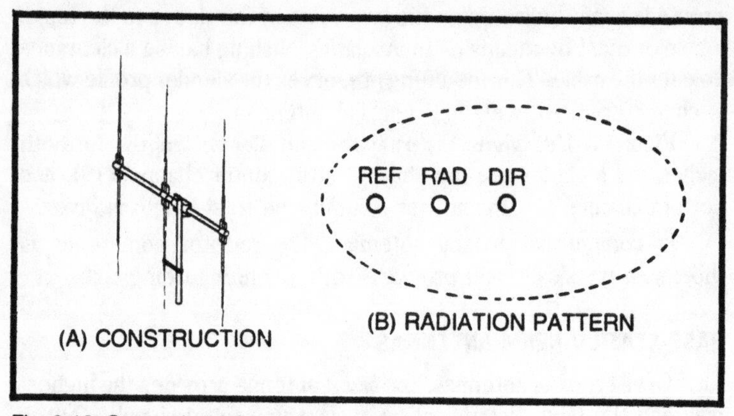

Fig. 3-16. Conventional three-element vertical beam.

one direction. For transmission in a particular direction, this antenna must be pointed in that direction, and the beam may be fixed or rotatable. Whenever there is room for its installation and operation, a rotatable beam antenna is a definite advantage at the CB base station. When communications always are in the same direction, as between two or more base stations, the beam antenna need not be rotatable, but may be permanently pointed in that direction.

CB beam antennas may be polarized either horizontally or vertically; but since most base stations communicate with mobile stations having vertical antennas, most CB beams are of the vertical type. These beams follow conventional design practice, but there is some variation in the models produced by different manufacturers.

The usual CB directional antenna contains a radiator element, a reflector element spaced behind the radiator, and one or more director elements spaced out in front of the radiator. In this arrangement, RF power is delivered only to the radiator. The reflector and directors are excited by the field of the radiator. The energy that they radiate is (because of the spacing between themselves and the radiator) in the correct phase to reinforce the radiated signal in the desired direction. Because the reflector and directors are excited electrostatically or electromagnetically, rather than by direct connection to the transmitter, they are called parasitic, and the assembly is known as a parasitic-type beam. The rudiments of parasitic beams are given in Chapter 2. Sections in that chapter may be consulted, keeping in mind that dimensions given therein are for amateur frequencies. Element lengths and spacings would need to be increased proportionately for the class D citizens band.

Three-Element Beam

One of the most popular CB directional antennas is the three-element beam. This antenna is manufactured in a number of variations. Figure 3-16A shows the basic model using telescoping elements for close adjustment of radiator, reflector, and director lengths.

From Fig. 3-16A, it is easily seen that this antenna is equipped with a gamma match for matching the lower impedance of the radiator to a 50-ohm coaxial line. (See Fig. 3-3B for a detail of the gamma match.) For impedance matching, a beat match (Fig. 3-3A) is sometimes used instead of the gamma.

Typical performance of this antenna is 8 to 9 dB forward gain, and 25 dB front-to-back ratio.

Multielement Parasitic Beams

In a parasitic beam, the gain and narrowness of the radiation pattern increase as the number of directors is increased. The number of elements can be increased up to the limit imposed by the dimensions of the boom required to support them, and the resulting weight of the assembly. Five and six elements do not seem unwieldy at class D frequencies (five-element beams are rated typically at 12 dB gain). For a discussion of multielement parasitic beams, see Chapter 2, keeping in mind that actual dimensions given there are for amateur frequencies. Element lengths and spacings would need to be increased proportionately for the class D citizens band.

Figure 3-17 shows a six-element vertical beam of the Yagi (parasitic) type. A typical version of this antenna is rated at 17 dB forward gain and 31 dB front-to-back ratio.

Figure 3-18 shows a special high-gain, eight-element parasitic antenna array (Wilson 500) in which the radiator and the six directors consist of crossed horizontal and vertical elements for both polarizations at the same time (this construction amounts to mounting two parasitic antennas—one horizontal, one vertical—on the same boom), and the reflector is a quad. (See Chapter 2 for an explanation

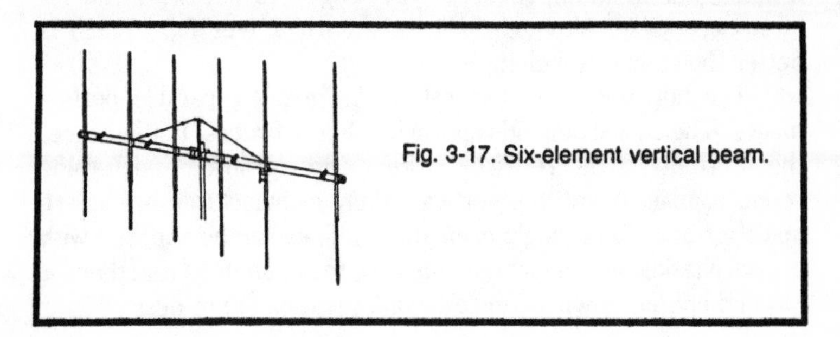

Fig. 3-17. Six-element vertical beam.

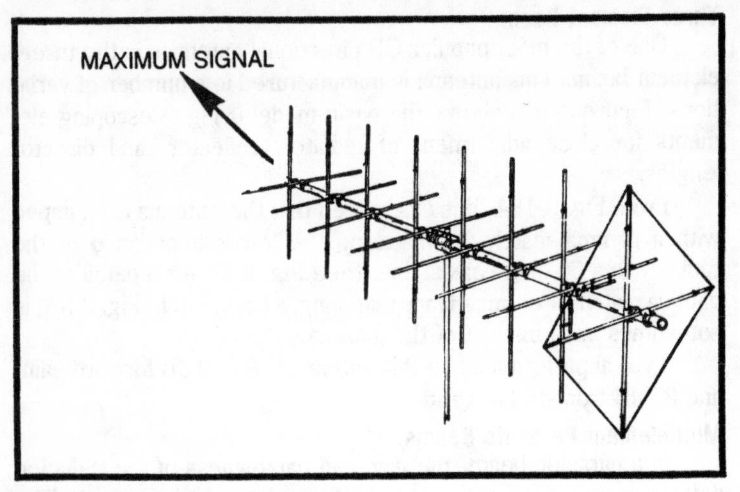

Fig. 3-18. Combination high-gain array.

of quad antennas.) This antenna requires a 40-foot boom, but is rated at 18 dB forward gain, 50 dB front-to-back ratio, and 50 dB side rejection.

Stacked Beams

Further improvement may be obtained by operating two identical beam antennas simultaneously side by side or one above the other. This arrangement is termed a *stacked array*. Figure 3-19 shows two stacked arrays—one, vertical (Fig. 3-19A) and the other, horizontal (Fig. 3-19B)—each consisting of two three-element parasitic beams. The stacked array gives an increase of approximately 3 dB over the gain of a single three-element beam and improves the front-to-back ratio.

The two beams in each array are fed simultaneously in phase from the same coaxial cable. The separation between them is best 1 wavelength (this is 36 feet 2.5 inches for channel 19, the center frequency of the band), but a half-wavelength (18 feet 1.25 inches) is often acceptable, although ⅝-wavelength (22 feet 7.5 inches) is better than a half-wavelength.

The horizontal and vertical stacks give comparable performance. The choice depends upon a number of factors. For instance, the horizontal array requires more mounting space and turning room, compared with the vertical. Many manufactured beams are made for use either singly or in stacks. They can be supplied with coaxial phasing harness so that it is easy to fasten them together for stacked operation when single-beam operation is not desired.

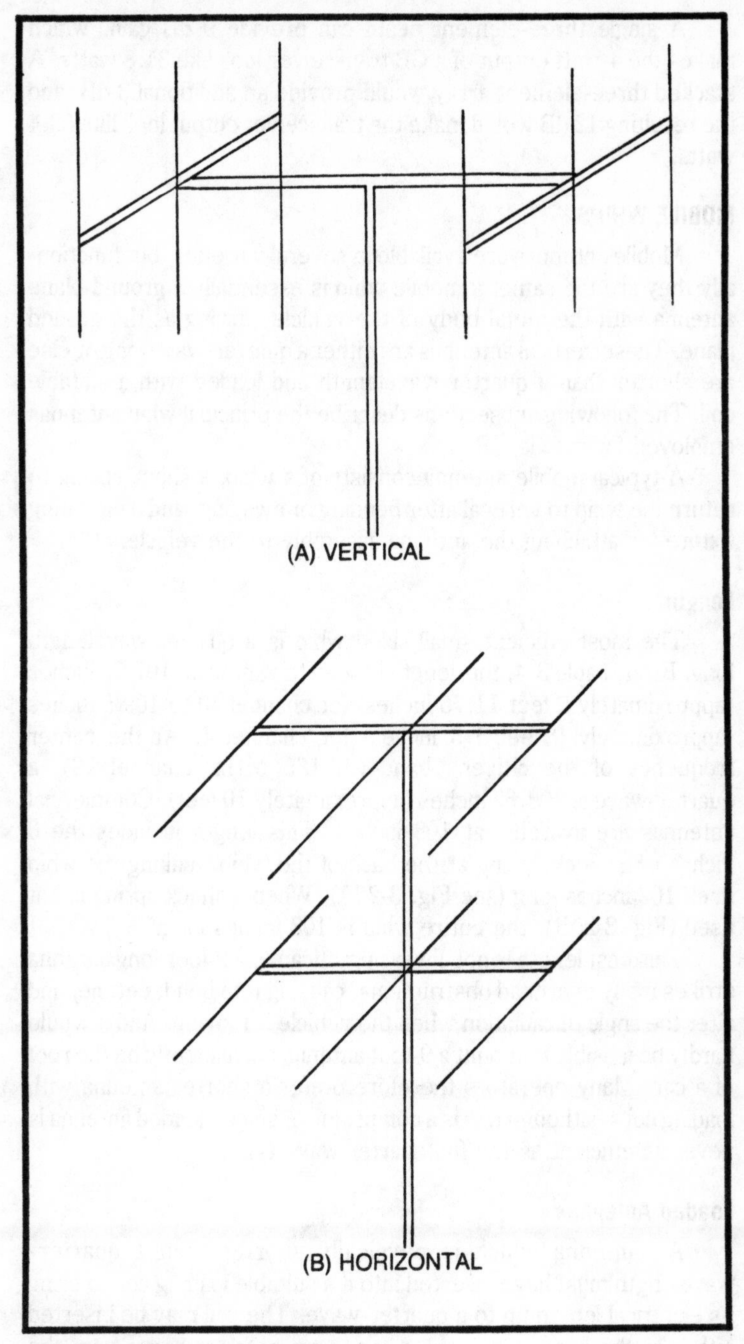

Fig. 3-19. Stacked beam antennas.

A single three-element beam can provide 9 dB gain, which makes the 4-watt output of a CB transceiver look like 31.8 watts. A stacked three-element array would provide an additional 3 dB, and the resulting 12 dB would make the transceiver output look like 63.4 watts.

MOBILE WHIPS

Mobile antennas are available in several varieties, but functionally they are the same: a mobile whip is essentially a ground-plane antenna with the metal body of the vehicle serving as the ground plane. These vertical antennas are either a quarter-wave long or else are shorter than a quarter-wavelength and loaded with a suitable coil. The following subsections describe the principal whip antennas employed for mobile CB.

A typical mobile antenna consists of a whip, a shock spring to return the whip to vertical after bending or swaying, and a mounting fixture for attaching the antenna assembly to the vehicle.

Length

The most efficient small-sized whip is a quarter-wavelength long. From Table 3-4, this length is seen to vary from 107.72 inches (approximately 8 feet 11.75 inches) for channel 40 to 109.5 inches (approximately 9 feet 1.5 inches) for channel 1. At the center frequency of the citizens band (27.185 MHz, channel 19), a quarter-wave is 108.62 inches (approximately 10 feet). Commercial antennas are available at 108 inches. This length includes the 6 inches of a shock spring at the base of the whip, making the whip itself 102 inches long (see Fig. 3-20A). When a shock spring is not used (Fig. 3-20B), the entire whip is 108 inches long.

This best length is not always practicable; a 9-foot-long antenna strikes many overhead obstructions, causing it to bend, detune, and alter the angle of radiation when the vehicle is moving. And it would hardly be feasible to mount a 9-foot antenna permanently on the roof of a car. Many operators therefore prefer a shorter antenna with loading coils, although this is a compromise, since a loaded antenna is never as efficient as the full quarter-wave type.

Loaded Antennas

An antenna which is physically shorter than a quarter-wavelength must have inserted into it a suitable loading coil to bring its electrical length up to a quarter-wave. The coil may be inserted either at the base, top, center, or three-quarters away from the

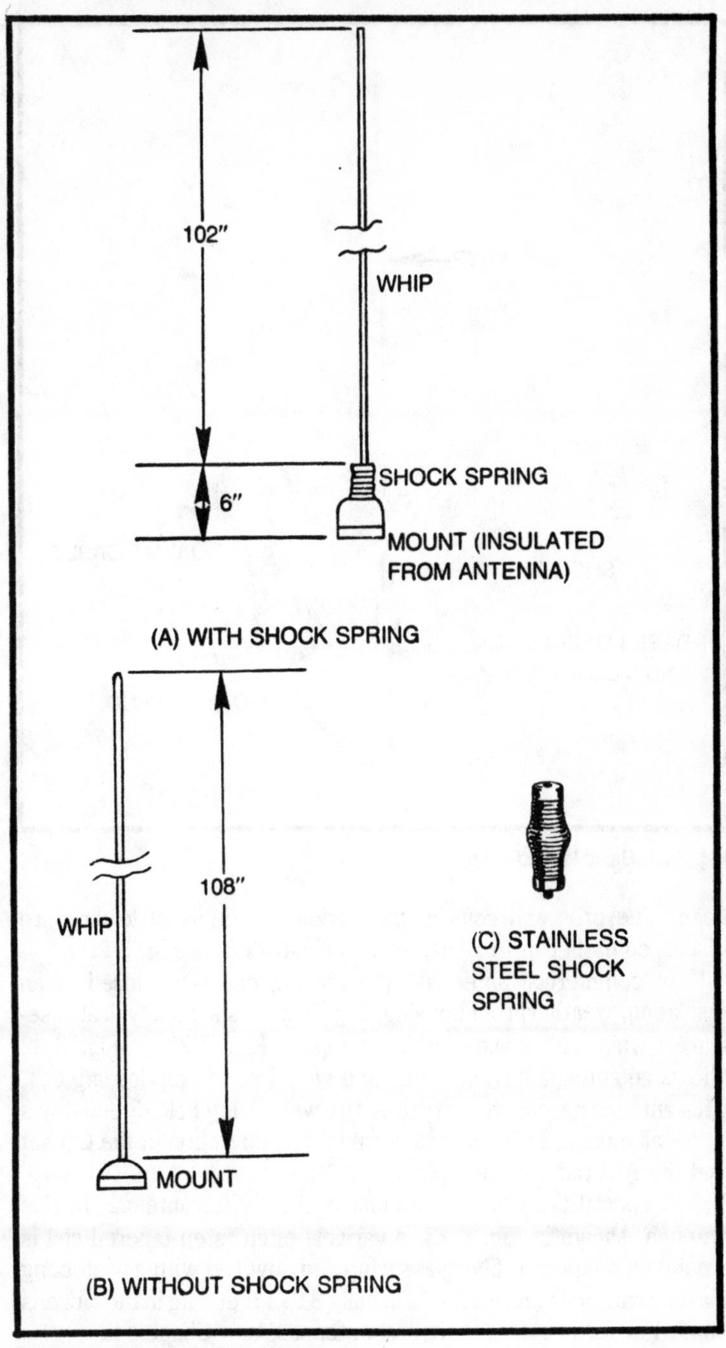

Fig. 3-20. Unloaded whip antenna.

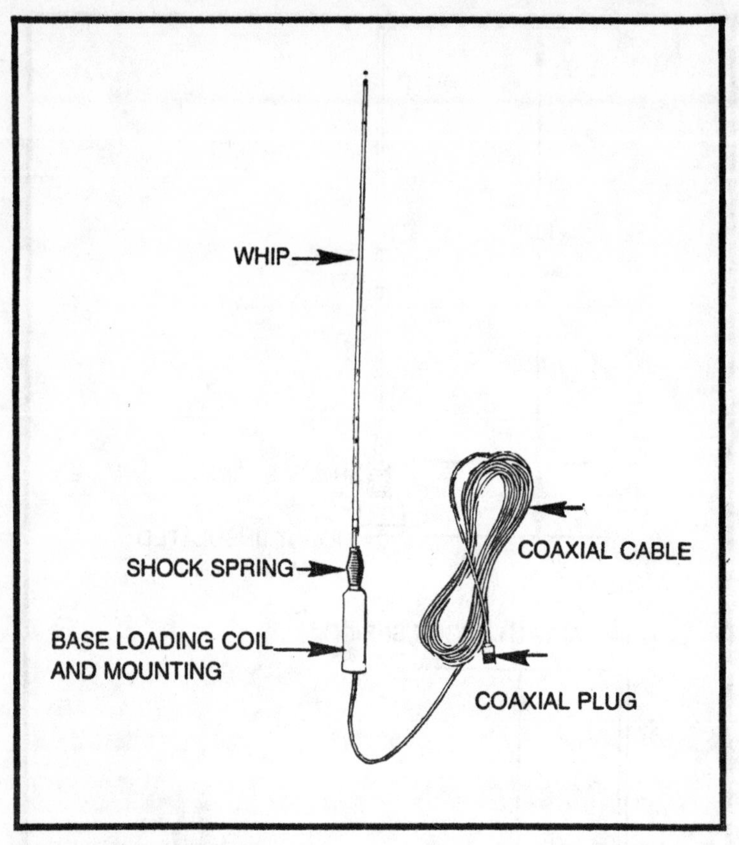

Fig. 3-21. Basic loaded whip.

base. The pros and cons of the various methods of loading are discussed earlier in this chapter and illustrated by Fig. 3-2.

In commercial antennas, the loading coil is enclosed in an insulating, weatherproof housing. Figure 3-21 shows a typical base-loaded whip with coaxial cable (compare Fig. 3-20A). Fig. 3-22 shows an antenna having a long and very slender top loading coil. This antenna has no shock spring, the whip being held directly by a split-ball mount. This antenna permits dual operation of the CB set and the AM radio in the car.

A special type of coil antenna is the *helical* antenna. In this antenna, shown in Fig. 3-23, a variable-pitch, step-tapered coil is wound on a tapered, fiberglass whip and covered with a protecting plastic coat. Such an antenna is usually 3 to 4 feet long in the citizens band. For further data on helical antennas, see Chapter 2.

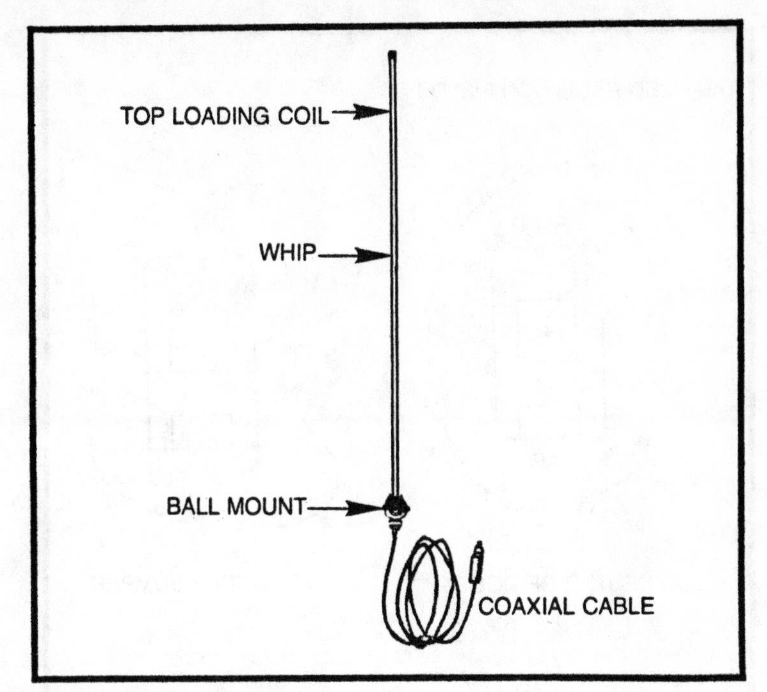

Fig. 3-22. Top-loaded whip.

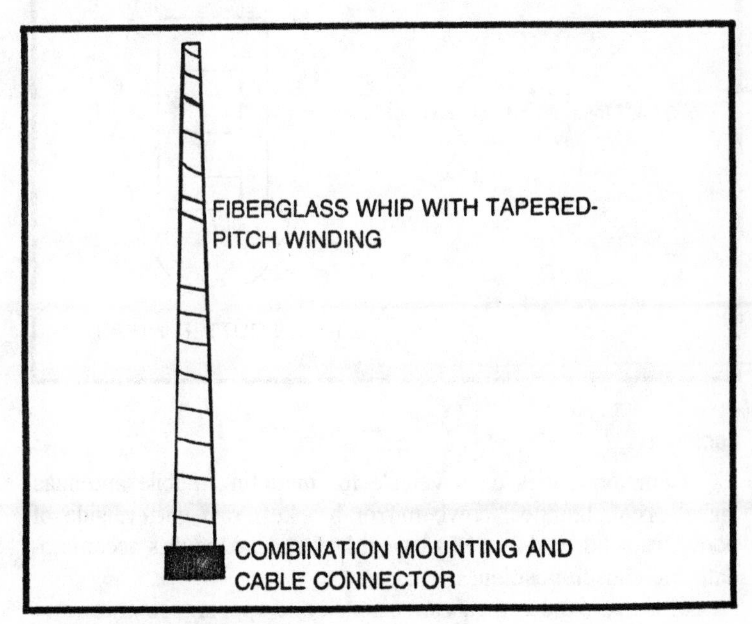

Fig. 3-23. Helical antenna.

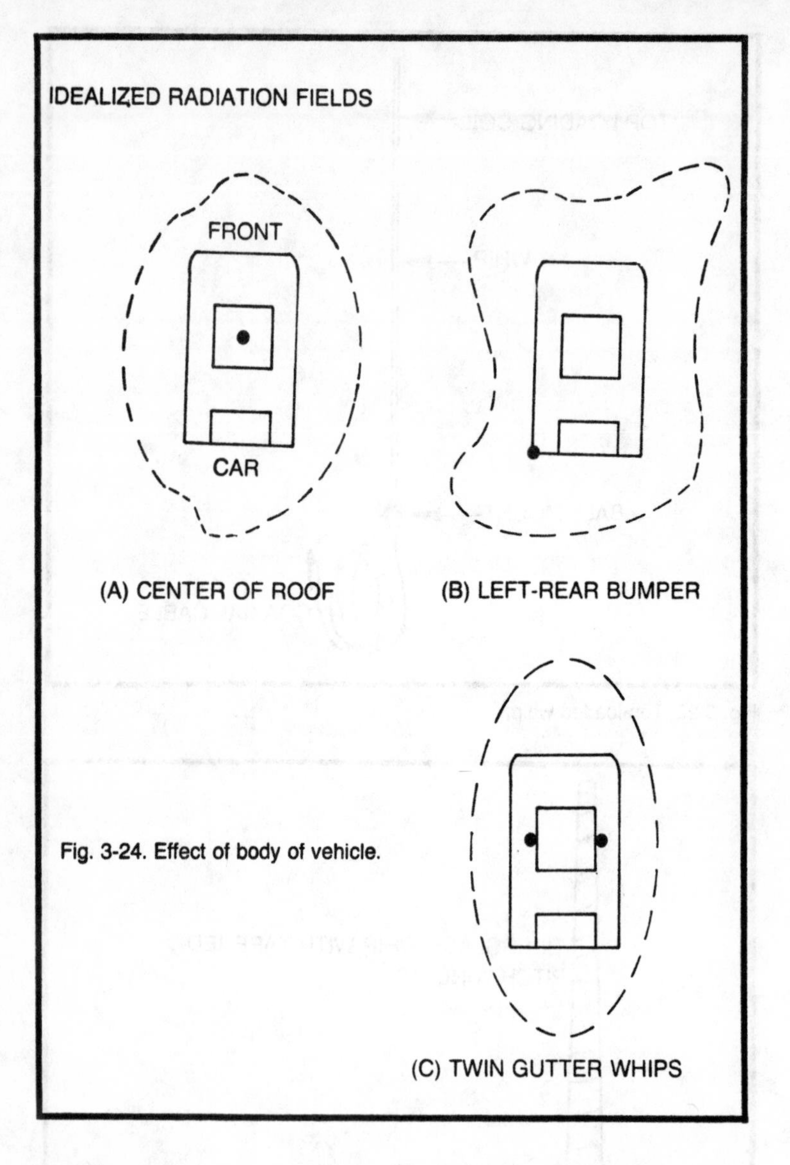

Fig. 3-24. Effect of body of vehicle.

Location

Common places on a vehicle for mounting mobile antennas include roof, bumper, cowl, mirror bracket, rain gutter, side of body, trunk lid, and fender. A wealth of special mounts accommodate the chosen position.

However, more than convenience and appearance enter into the selection of an antenna location. The metal body of a vehicle distorts the normally circular radiation pattern of the vertical antenna. Thus, it makes a difference where on the car the antenna is installed. Figure 3-24 gives some idealized radiation patterns corresponding to antennas mounted in the center of the roof (Fig. 3-24A), and on the left-rear bumper (Fig. 3-24B). Figure 3-24C shows how a down-the-road directional pattern may be obtained with two identical antennas mounted on opposite rain gutters and fed in phase with a suitable coaxial harness. The latter scheme is often employed by truckers and drivers of recreational vehicles who install the antennas on opposite mirror brackets. Figure 3-25 shows commercial-type whips and phasing (diplexer) harnesses for this purpose; Fig. 3-25A shows straight whips, and Fig. 3-25B shows center-loaded whips. *Cophased* antennas of this type are used also on the left and right ends of a rear bumper.

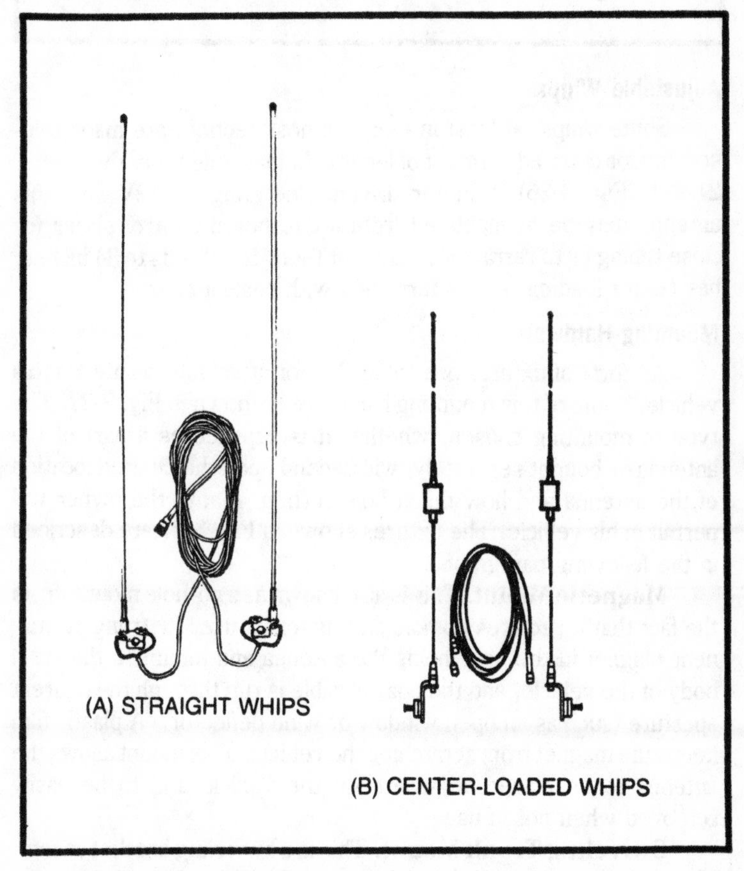

Fig. 3-25. Double-trucker antenna.

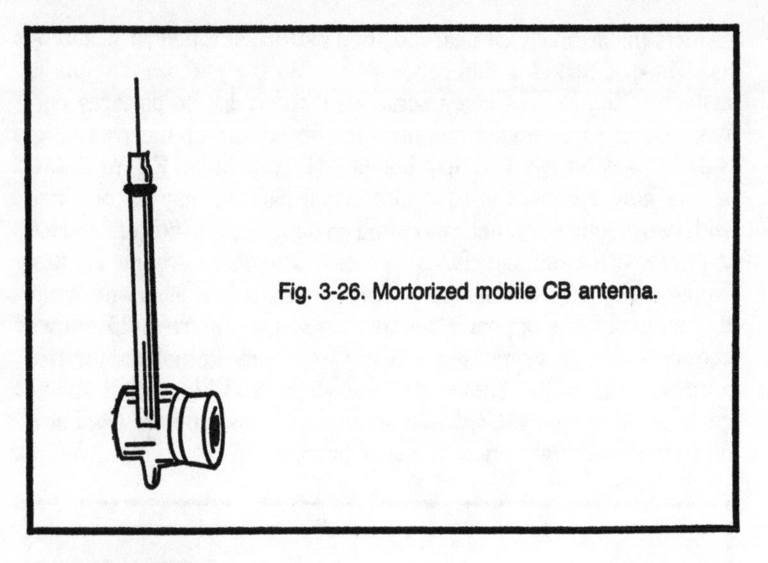

Fig. 3-26. Mortorized mobile CB antenna.

Adjustable Whips

Some whips, at least in their topmost section, are made telescoping for close adjustment of length. At least one type, the Archer 21-971 (Fig. 3-26), is motor driven. Operating on 12V DC, this antenna may be manipulated from a dashboard control either for close tuning or to retract it to prevent theft. It extends to 34 inches, has center loading, and is furnished with coaxial cable.

Mounting Hardware

All sorts of fixtures are obtainable for attaching an antenna to a vehicle. Some of this mounting hardware is shown in Fig. 3-27. The type of mounting chosen, whether it is supplied as a part of the antenna or bought separately, will depend upon the desired location of the antenna and how much hole cutting, if any, the owner will permit in his vehicle. The fixtures shown in Fig. 3-27 are described in the following paragraphs.

Magnetic Mount. This is also known as a no-hole mount, from the fact that it requires no hole cutting for its use. A strong permanent magnet in the base holds the antenna and mount to the steel body of the vehicle, and the coaxial cable is run through the nearest aperture (such as an open window or wind deflector). A plastic film keeps the magnet from scratching the vehicle. This mount allows the antenna to be placed anywhere on the vehicle and to be easily removed when not in use.

Swiveling Trunk Mount. This is a swiveling bracket assembly which attaches to the lip of the car trunk. When the antenna is not in use, this swivel action allows the antenna to be folded back into the trunk.

Screw-On Gutter Mount. This is a screw-on type of mounting which requires only two screws and no hole cutting to hold the antenna to the rain gutter of a car. Note also the gutter mounts on the antennas in Fig. 3-25B.

Spring Gutter Mount. This fixture, unlike the gutter mount in Fig. 3-27C, requires neither screws nor holes, but has a thumb-operated spring clamp to fasten it to the rain gutter of a car.

Gutter Clip. This is a mount which clips into the rain gutter of a car and is secured to the gutter by tightening a single screw.

Bumper Mount. This is a stainless steel strap which fastens around the bumper and holds the base of the antenna. Figure 3-27G shows how this is accomplished. Other bumper mounts use a chain instead of a strap.

Bumper-Mounted Antenna. Here, an antenna is shown secured to a bumper by means of the strap-type mount shown in Fig. 3-27F.

Ball Mount. This assembly consists of a half-ball which is fastened to the base, and a full-ball which rotates within the half-ball and carries the antenna. This mounting allows the antenna to be set in the vertical position regardless of the position in which the base must be attached to the vehicle. Thus, a ball mount is usable on any conceivable surface of the vehicle—flat, curved, horizontal, or vertical. Figure 3-27M also shows a ball-type fixture.

Cowl Mount. This fixture, which requires a mounting hole up to a 1-inch diameter, depending upon make and model, usually holds the antenna to a front or rear fender or other curved surface. The antenna is screwed into the mount.

Stationary Trunk Mount. This is a metal bracket (Fig. 3-27B shows a more complicated such bracket) enclosed in a white cover. Tightening two screws in the assembly fastens the bracket to the lid of the trunk. A 0.75-inch-diameter hole is required.

Mirror Mount. This unit fastens a rooftop-size antenna to a mirror mount (strut) or station wagon rack. See, for example, the mirror mounts on the antennas in Fig. 3-25A.

Snap-In Rooftop Mount. This fixture (Fig. 3-27L) accommodates an antenna to the roof of the vehicle. It requires a 0.375-inch-diameter hole and snaps together without screws.

Body Mount. This is a ball-joint fixture (Fig. 3-27M) which allows the antenna to be mounted on any vertical portion of the vehicle. It requires screw holes (three or four, depending upon make and model), as well as a clearance hole for the coaxial cable.

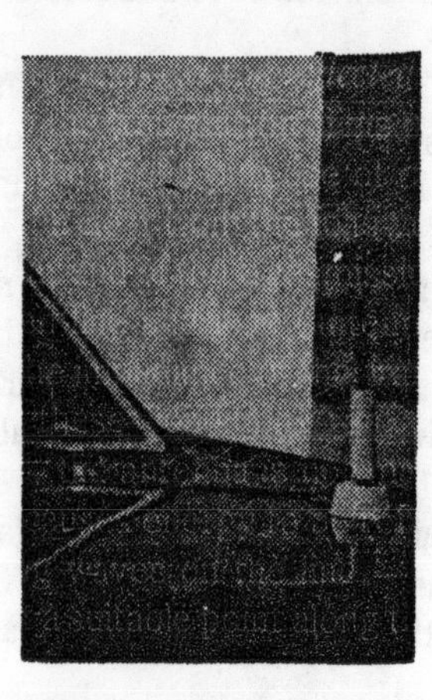

(A) MAGNETIC MOUNT

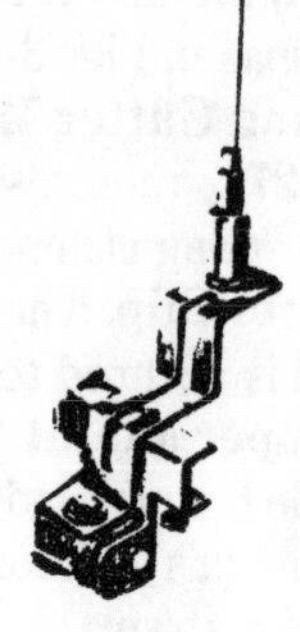

(B) TRUNK MOUNT

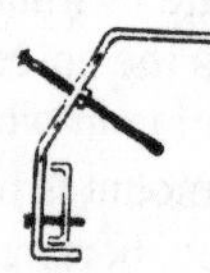

(C) GUTTER MOUNT
(SCREW TYPE)

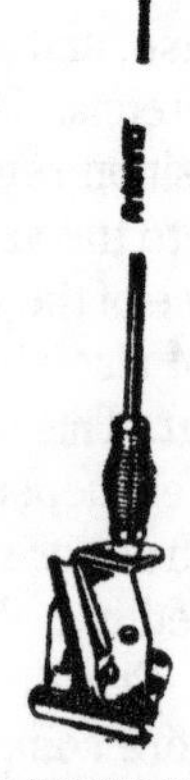

(D) GUTTER MOUNT
(SPRING-CLIP TYPE)

(E) GUTTER CLIP

(F) BUMPER MOUNT

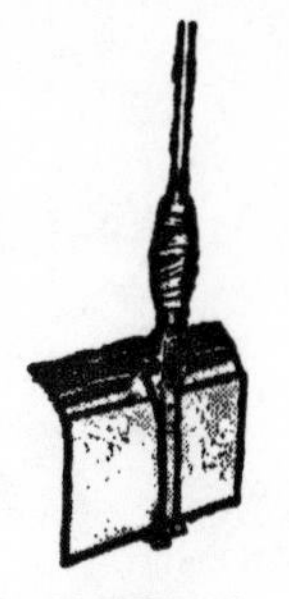

(G) ANTENNA TIED TO BUMPER

(J) TRUNK MOUNT

(H) BALLMOUNT

(I) COWL MOUNT

(K) MIRROR MOUNT

(L) SNAP-IN ROOF-TOP MOUNT

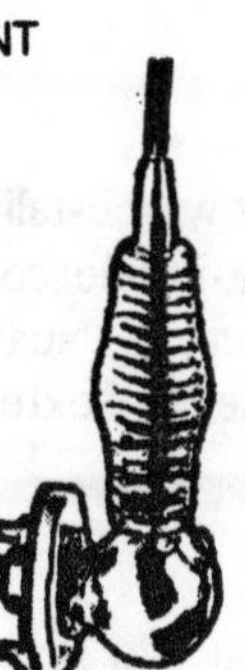

(M) BODY MOUNT

Fig. 3-27. Mounting hardware for mobile whips.

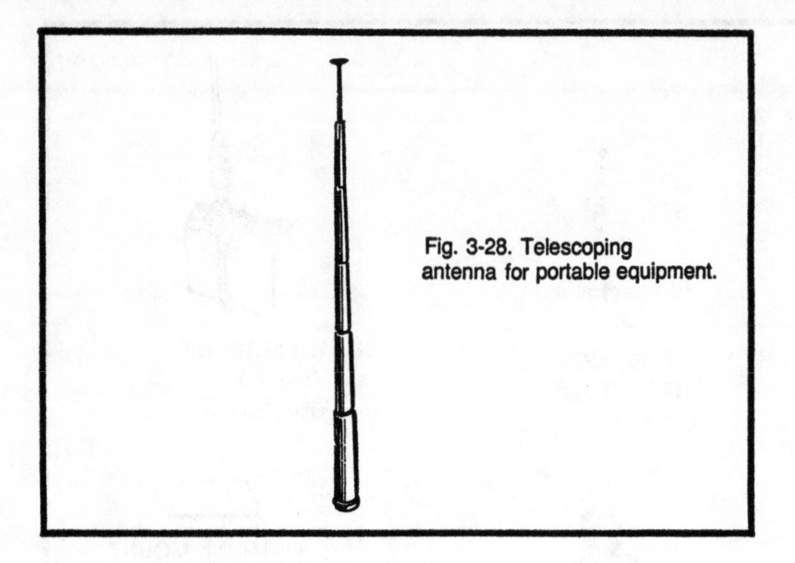

Fig. 3-28. Telescoping antenna for portable equipment.

MISCELLANEOUS ANTENNAS

Occasionally, a CB set must be operated with something other than a standard antenna. For this purpose, a wide variety of devices have been used—temporarily or permanently—with various degrees of success. Some of the more meritorious of these are described below.

SPECIAL PRECAUTION

Many emergency and makeshift antennas—even those described here—do not provide a good impedance match to the transmitter. Always check SWR with your transmitter before putting such an antenna into use. This will prevent damage to the transmitter.

Integral Antenna

In midget CB transceivers of the walkie-talkie type and in some desk-top types, the antenna is a built-in telescoping whip (see Fig. 3-28) extending from the top of the case. Usually, antennas of this type collapse all the way into the case, and extend outward to a full length of about 36 inches.

All-Purpose Midget Whip

A small, telescoping whip, such as that shown in Fig. 3-28, may be bought as a separate component that extends to 36 inches and collapses down to 6 or 7 inches. In some temporary installations, especially indoors, this antenna—often stood on top of the transceiver by means of a suction cup—is connected directly to the antenna terminal of the instrument. Fair CB communication over a relatively short distance is obtainable with this setup. However, it should be noted that this antenna when extended to its full 36 inches has an electrical length of only about 1/12-wavelength at 27 MHz.

Rabbit-Ears Antenna

This telescoping antenna (Fig. 3-29) is familiar as an accessory for TV and FM receivers. It consists of two telescoping legs mounted on a hinged joint which allows the angle of separation between them to be smoothly varied. This antenna may also be rotated. It is a dipole, with a short length of cable connected to the base of the legs.

The rabbit-ears antenna offers considerable flexibility, since the length of each leg may be adjusted independently of the other (usually up to 35 or 45 inches, depending upon make and model); the angle between the legs is adjustable almost from 0° to 180°; and the antenna is rotatable. For CB, it gives fair results over a relatively short distance. One must note, however, that this antenna when extended to its full 45 inches per leg has an electrical length of only about 1/5-wavelength at 27 MHz. (Some rabbit-ears have base loading coils and may be adjustable to a quarter-wavelength.) The

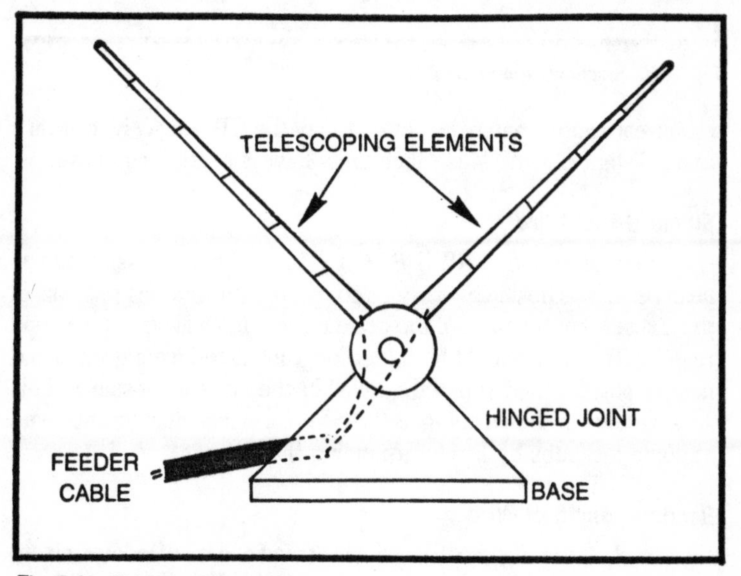

Fig. 3-29. Rabbit-ears antenna.

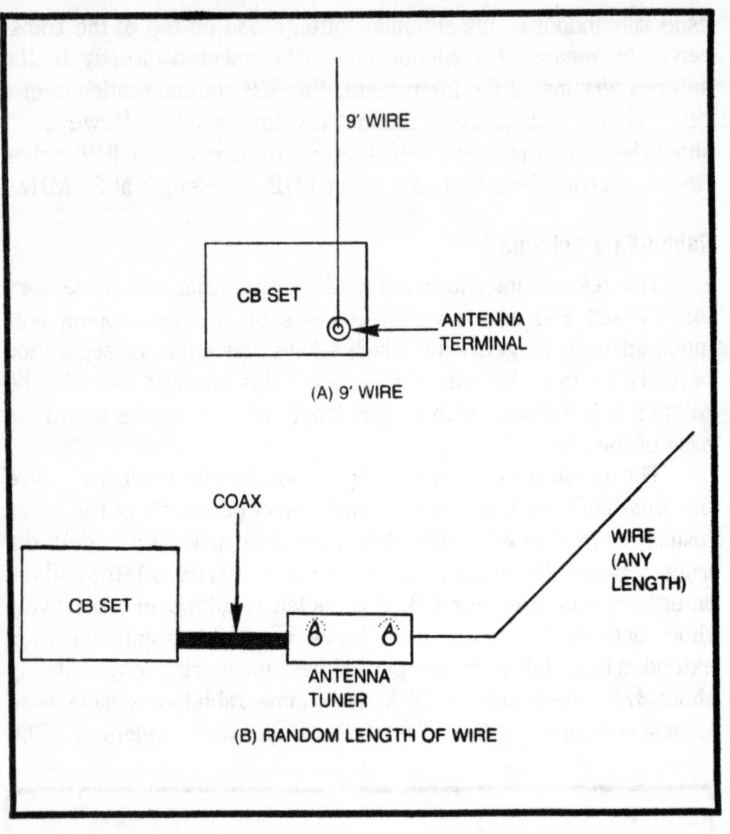

Fig. 3-30. Single-wire antennas.

rabbit-ears antenna may be set on top of the CB transceiver or on a more distant object. Most such units have a suction-cup base.

Single 9-Foot Wire

A single strand of AWG #12 or #14 solid wire—suspended as near vertical as possible—may be connected the antenna terminal of the CB set (hot terminal of the coaxial plug) and will often give good results. The wire should be 9 feet long, insulated from ground and nearby objects, and situated as well in the clear as possible. This arrangement is shown in Fig. 3-30A. Performance may be improved at some locations by firmly grounding the CB set.

Random Length of Wire

As shown in Fig. 3-30B, a single strand of wire of indeterminate length may function as an antenna if it is coupled to the CB set through a small antenna tuner (Archer Model 21-513, for example, matches the transceiver to any antenna impedance between 10 and 10,000 ohms). The wire should be insulated from ground and nearby objects, situated as well in the clear as possible, and suspended as near vertical as possible. The wire may be bent if this is unavoidable.

Indoor and Limited-Space Antennas

The foregoing antennas described are usable indoors and also semi-outdoors (on an outside windowsill, balcony, etc.). In such use, the antenna should be as well in the clear as possible and will give best performance when it is located on one of the upper floors of the building (the topmost floor usually is best). When a ground connection is needed, it should consist of a connection to the cold-water pipe of the building plumbing system. A base-, center-, or top-loaded cartop antenna is short enough to stand on top of the transceiver or on the operating desk.

Chapter 4
Shortwave-Listener Antennas

Shortwave listeners (SWLs) listen to a great many radio services over a wide frequency range. These include marine, foreign and domestic shortwave broadcast, amateur, police, fire, aviation, CB, weather, standard time and standard frequency, and many more. Some SWLs even listen to radio transmissions below the standard broadcast band (that is, below 500 kHz), in addition to actual *short* wave signals. Such hobbyists perhaps should be called *all-wave listeners*, instead of SWLs. In this chapter, however, we are concerned with listeners to the short waves, that is, to frequencies higher than 1600 kHz. At present, nobody knows the exact number of SWLs. But it is known that they are very numerous and that many of them use nondescript antennas.

SWLs legally eavesdrop on all frequencies; few confine their listening activity to a single band or service. Because of this wide frequency coverage, compromise operation of the antenna usually is necessary. Since SWLs do not transmit, however, their antennas need not be heavy duty (power type).

The antennas described in this chapter have been found to operate well in SWL service. They can be altered and modified in various ways to suit individual demands of installation and performance.

SPECIAL REQUIREMENTS

What makes a good SWL antenna? An activity, such as shortwave listening, that covers so many frequencies, radio services, and signal strengths demands versatility in an antenna. A good SWL antenna, thus, first of all picks up the strongest signal and the least noise—natural and manmade—over the widest frequency range. When most of the listening takes place in one band, the antenna may be designed for optimum performance in that band. But when listening is in several bands, the antenna ideally will work best near the center or lower end of the desired spectrum and will give good compromise operation at other points.

The following paragraphs contain 11 factors which the SWL should pay attention to when thinking about an antenna.

Matching

An efficient antenna is a matched system; that is, the antenna impedance, transmission line (feeder) impedance, and receiver input impedance are all equal. Under these conditions, the maximum signal power is transferred from the antenna to the receiver and the most efficient reception results.

For wide-band operation (for example, 1.6–30 MHz), it is difficult to maintain this impedance match throughout the range. In most such cases, the system is designed for the best match either at the lowest frequency or at the center of the range, and the resulting compromise is accepted at all other frequencies. If any impedances in the antenna system (antenna, feeder, receiver input) do not naturally match, an impedance-matching device—such as a transformer or tuner—must be employed. Common all-wave-receiver input impedances are 50, 75, and 300 ohms. A few of the older receivers have 600-ohm inputs.

Frequency Coverage

The typical SWL operates between 1.6 and 30 MHz, the tuning range of the average all-wave receiver. A few hobbyists extend their listening into the spectrum above 30 MHz, which includes amateur FM, TV, and other services. Most SWLs will prefer to use one antenna from 1.6 to 30 MHz and will employ special antennas at the higher frequencies. When an SWL is interested in only a part of the spectrum (for example, two ham bands), he may prefer to use a special antenna for the bands of interest.

Receiver Sensitivity

All-wave receivers are sensitive instruments. The better ones, for example, deliver full rated audio output for a few microvolts of RF signal input. Because of this feature, some SWLs feel that receiver sensitivity will compensate for a bad antenna, and that "any old piece of wire will do." While it is true that a sensitive receiver may adequately handle the low-signal output of a compromise antenna, it is true also that an overworked receiver will pick up more noise (which may mask the signal). High receiver sensitivity, while certainly a good thing to have, is no excuse for failure to provide the best antenna possible at a given location.

Outside vs Inside

Generally, an outside antenna is preferable to an inside one. Sometimes, however, as in some apartment houses and office buildings, outside antennas are prohibited; but this is not a complete catastrophe, as most sensitive all-wave receivers work quite well with an inside antenna, except perhaps on the ground floor of some metal-frame buildings.

An inside antenna must be situated as far as possible from metallic objects. Also, one must not asume that because the antenna is inside a nonmetallic building it will not be shielded, for some buildings—such as the stucco variety—may have a shell of metal mesh in their structure.

Horizontal vs Vertical

The SWL must choose between horizontal and vertical orientation of the antenna after taking several factors into consideration. (1) The vertical antenna takes the lesser horizontal space. (2) The vertical antenna is omnidirectional; that is, receives from all directions approximately equally well. (3) The vertical can be made free standing if it is constructed of rod or tubing. (4) The horizontal is somewhat directional, receiving best off its center and least off its ends (see, for example, Fig. 4-1A). (5) The horizontal takes the greater horizontal space. (6) The vertical picks up the more electrical noise, including static. (7) The horizontal picks up the lesser noise.

Height

The antenna should be erected as high as practicable, to get it in the clear away from ground, buildings, trees, and other objects, both conducting and nonconducting. In working for maximum height, however, the erector must remember the height above ground affects the impedance of an antenna (see Fig. 1-10 in Chapter 1). Any laws, regulations, or ordinances pertaining to antenna height must be obeyed.

Length

Ideally, an SWL antenna should be a quarter-wave or half-wave (or some multiple of a quarter-wave) long at the frequency at which the antenna will most often be used. When the antenna is to be used regularly at a number of frequencies, this will be the lowest frequency at which it will be used. The length of wire for a half-wave antenna may be determined with the following equation:

$$l = 468/f \qquad (4\text{-}1)$$

where l = length in feet, and
f = frequency in megahertz.

Thus, a half-wave antenna might be designed for 1.6 MHz operation and would give compromise operation up to 30 MHz: $l = 468/1.6 = 292$ feet 6 inches.

In many instances, especially when the antenna is in a congested city, horizontal length is severely limited. In such case, the SWL might like to know the lowest frequency which is a half-wave at the maximum length allowed to him. Suppose, for example, the maximum antenna length possible at a given location is 35 feet. Then, the frequency at which this length of wire is a half-wave may be determined by rewriting Equation 4-1: $f = 468/l = 468/35 = 13.37$ MHz.

Directivity

The type of antenna chosen will often depend upon the points of the compass from which the SWL wishes to receive signals. The vertical antenna is omnidirectional, receiving from all directions approximately equally (see Fig. 4-1D). The horizontal half-wave antenna (Fig. 4-1A), on the other hand, receives poorly (almost not at all) off its ends and best off its center. For the half-wave horizontal, this results in a doughnut pattern. As the horizontal antenna is made longer than a half-wave, more major and minor intensity lobes appear off the centers, as shown in Figs. 4-1B and 4-1C.

From this, it is seen that a United States short-wave listener favoring European broadcasts would use a north-south horizontal antenna, whereas a United States SWL wanting reception from South America would use a generally east-west horizontal. And an SWL desiring universal reception would use a vertical.

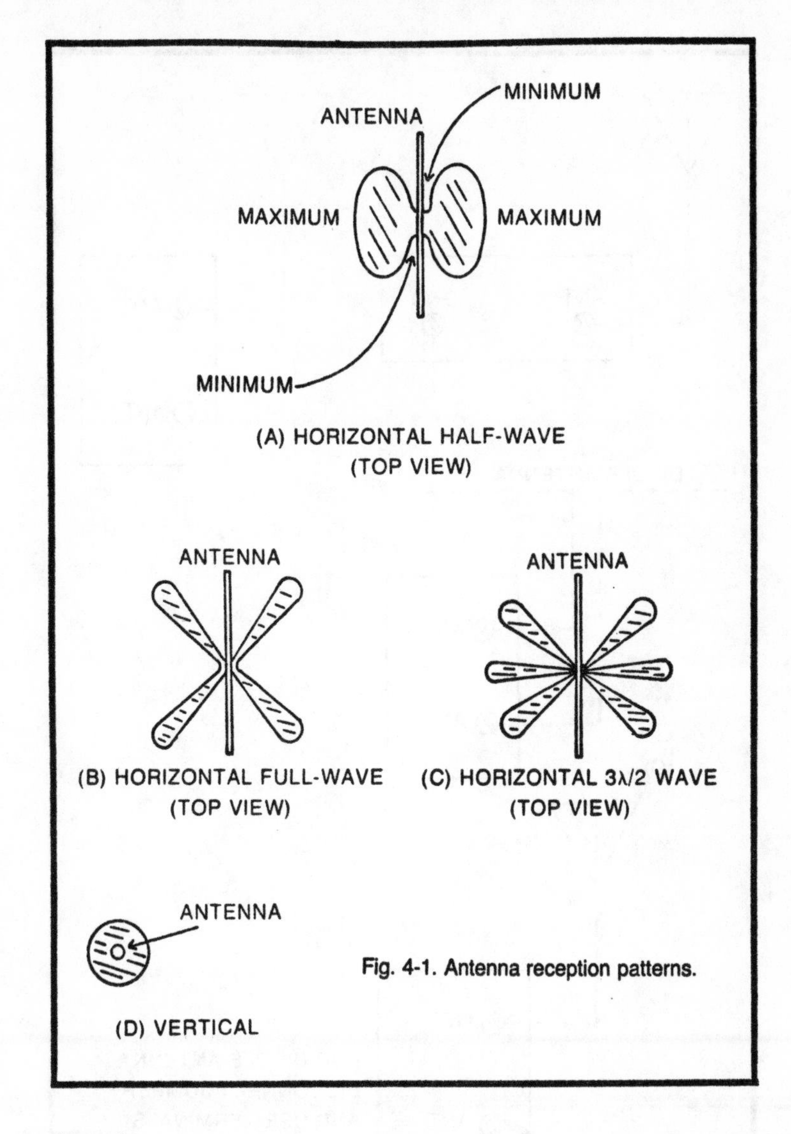

Fig. 4-1. Antenna reception patterns.

Ground

SWL antenna operation sometimes is enhanced by grounding the receiver. The ground connection need not be elaborate; usually, connection to a cold-water pipe will suffice.

Sometimes, one of the input terminals of the receiver is a ground terminal (marked GND, as in Fig. 4-2A). In other instances, two *antenna* terminals are provided (as in Fig. 4-2B), but one of these can be grounded. Whenever a ground is used, the connections

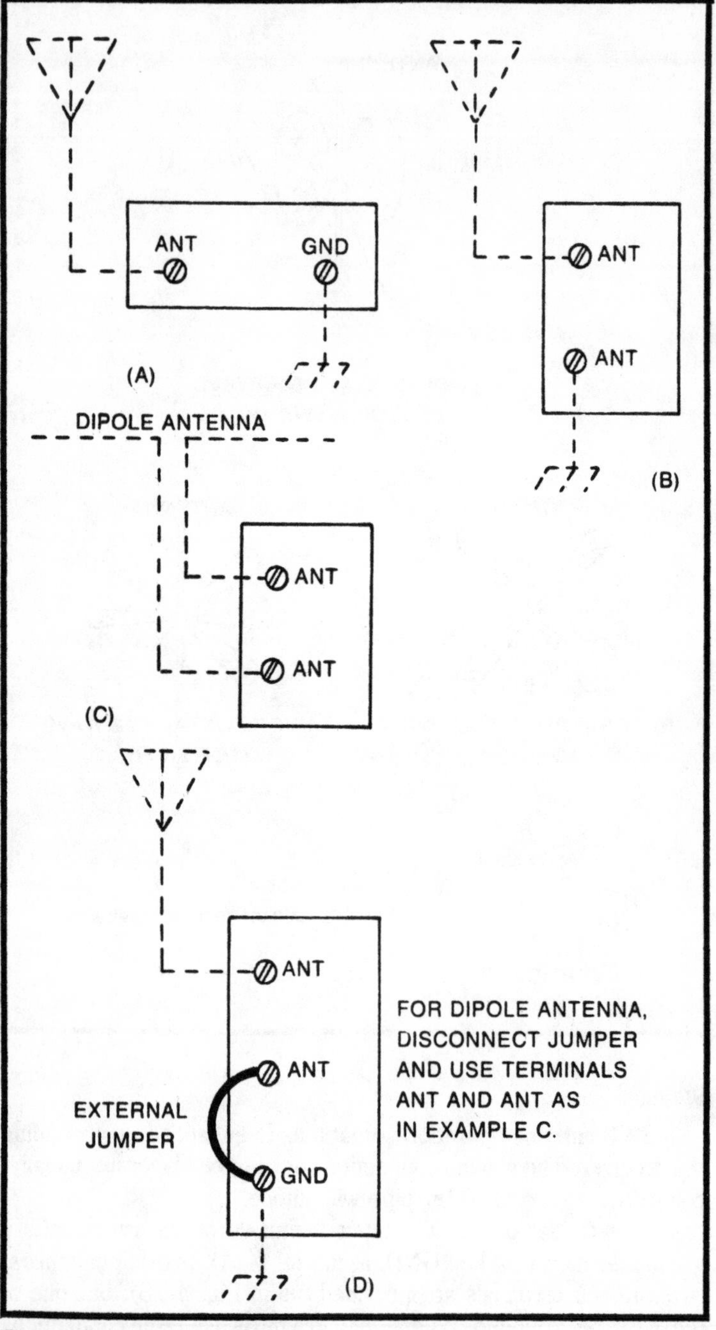

Fig. 4-2. Receiver input terminals.

to the receiver antenna input terminals should be swapped for loudest signal.

Location

The SWL antenna should be as close as practicable to the receiver, yet as high as practicable in the air—a difficult double requirement to meet. An efficient transmission line will allow the antenna to be situated as high as is needed, and still give operation that approximates having the receiver right at the antenna.

Regarding location of the antenna itself, it is well to remember that an antenna picks up not only radio signals, but also electrical noise generated by power lines, machinery, sparking contacts, and other sources. As much signal and as little noise as possible should be fed into the receiver by the antenna. For noise-free operation, erect the antenna as far as practicable from power lines and electrical machinery. Run a horizontal antenna wire at right angles to the direction of the power line. In cases of extreme interference, an auxiliary noise antenna can be employed to cancel noise signals in the main antenna.

Safety

For protection of the antenna, associated radio equipment, buildings, and persons, an antenna must be erected solidly. Furthermore, it should not be located where it might come into contact with power lines, telephone wires, or TV cables under any circumstances. The antenna should be provided with a lightning arrester or lightning switch. Any local ordinances pertaining to antennas or related elevated structures must be obeyed.

SIMPLE SINGLE-WIRE ANTENNAS

A single wire run straightaway from the antenna terminal of an all-wave receiver is the simplest possible antenna, since it requires no transmssion line, lead-in, or feeder. This antenna works very well when it can be used, but it is practicable only when the receiver is located at the top of a building, as on the top floor or in a penthouse, and when the wire can be brought directly into the room and run for a short distance to the receiver without bending.

When the receiver is located on the top floor of a building, the wire can be run out through a window or wall and through open space to another building or to a chimney, tree, post, or other object. But when the receiver is on a lower floor, the wire will usually be too close to ground and obstacles. Thus, the top floor is next in desirability to a penthouse, and the lower floors are progressively undesirable.

Horizontal

The most common form of simple single-wire antenna is the horizontal type (Fig. 4-3A). This consists of a single strand of AWG #12 or #14 solid copper wire. Where there is no choice in the matter, the wire may be any length that can be fitted into the available space. For best results, however, the wire length should correspond to an electrical half-wave (Equuation 4-1) or some multiple of a quarter-wave at the *most-used* frequency. As much of the wire as practicable should be outside of the building and situated well away from all objects.

A good-grade strain insulator or egg insulator must be attached to the far end to insulate the wire from the support. The near end of the wire may be brought through a porcelain-tube insulator passed through the wall of the building, or it may be attached to a suitable standoff insulator or feedthrough insulator. A flat, insulated lead-in strip also will allow the antenna to be passed under a closed window.

Inclined

When the far support must unavoidably be higher than the point at which the wire enters the building, the simple single-wire antenna will be tilted, as shown in Fig. 4-3B. All of the descriptive remarks regarding installation of the horizontal antenna in the foregoing subsection apply equally well to the inclined antenna and will not be repeated here.

Vertical

In those instances when the antenna wire can be run straight up from the receiver and through the roof of the building, the vertical version (Fig. 4-3C) may be used. This version of the simple single-wire antenna has the advantage afforded by all vertical antennas—omnidirectivity. A disadvantage of the arrangement, however, is its length restriction; that is, the need for a top support limits the wire to relatively moderate length. A tubing or rod version may be made somewhat longer, since it can be made self-supporting or guyed. There are instances in which an SWL has run a vertical wire down the side of a tall building or between buildings; but this is not a recommended arrangement, as the antenna pickup is obstructed by the buildings.

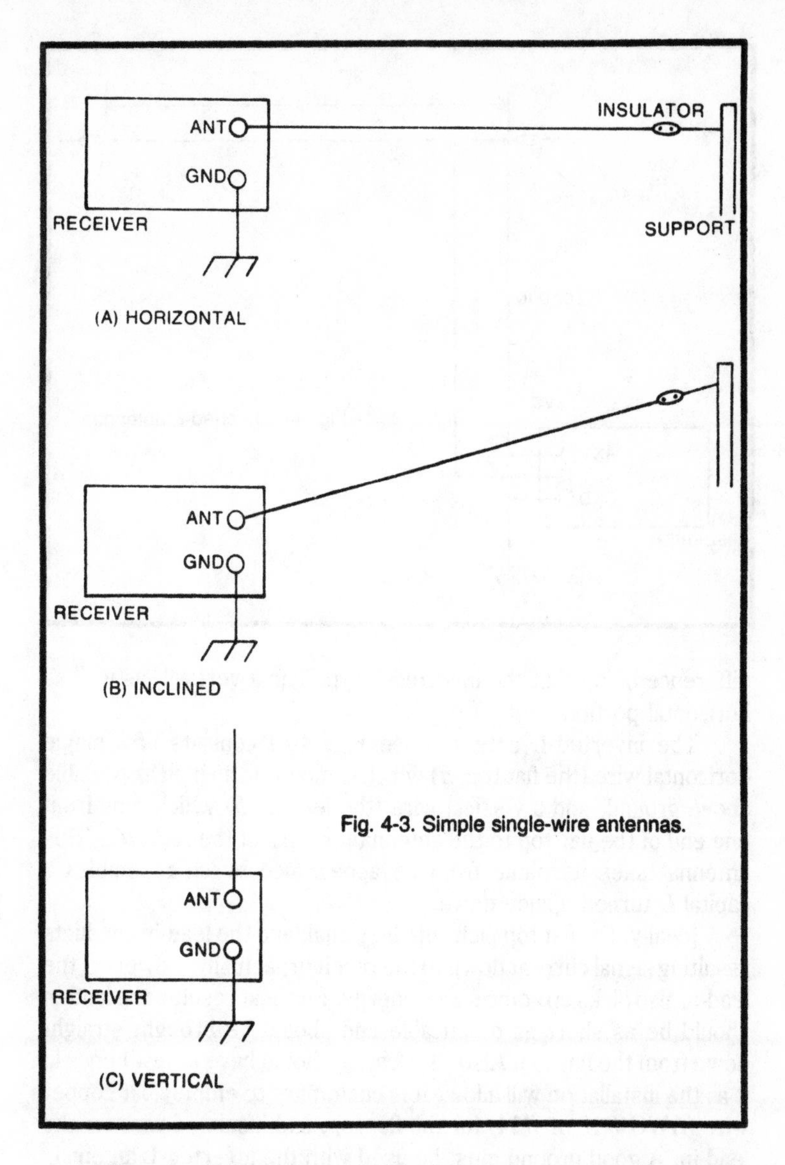

Fig. 4-3. Simple single-wire antennas.

When the SWL activity is limited to higher frequencies, a rigid vertical antenna, such as a ⅝-wave citizens band antenna, can be used (see Chapter 3).

INVERTED-L ANTENNA

Next to the simple single-wire antenna in popularity is the *inverted-L* antenna. These two antennas resemble each other, the

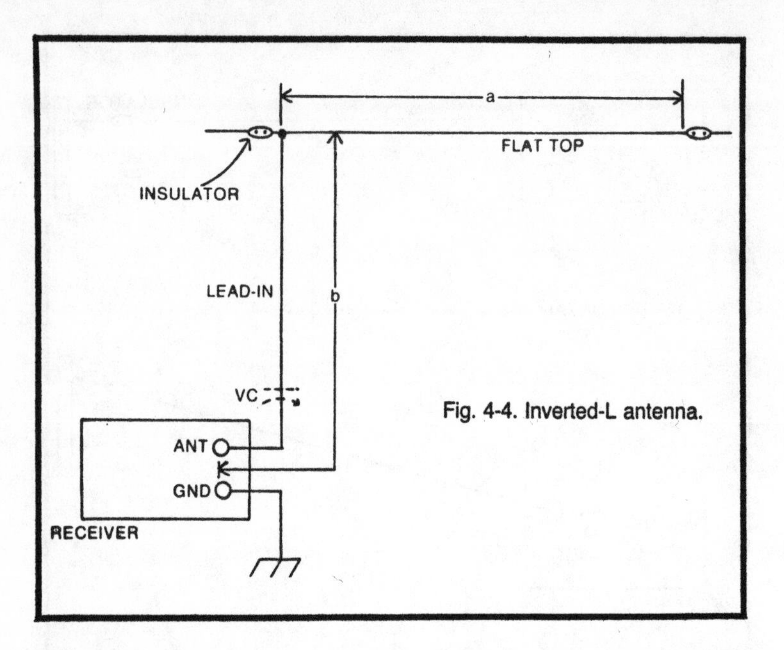

Fig. 4-4. Inverted-L antenna.

difference being that the inverted-L type has a vertical, as well as horizontal portion.

The inverted-L antenna (see Fig. 4-4) consists of a single horizontal wire (the flat top, *a*) which is mounted as high as possible above ground, and a vertical wire (the lead-in, *b*) which runs from one end of the flat top to the antenna terminal of the receiver. This antenna takes its name from its appearance which resembles a capital *L* turned upside down.

Ideally, the flat top picks up the signal, and the lead-in conducts resulting signal current down to the receiver; actually, however, the lead-in also picks up some signal energy. For best results, the lead-in should be as short as practicable and should be brought straight down from the flat top. Also, the lead-in should have as few bends in it as the installation will allow. It is customary to employ bar copper wire (AWG #12 or #14) for the flat top, and insulated wire for the lead-in. A good ground must be used with the inverted-L antenna.

Most SWLs who use the inverted-L antenna make the flat top as long as the available space at the location will permit. When most of the listening is done in one band of frequencies, however, the length of the flat top plus lead-in ($a + b$ in Fig. 4-4) should be a quarter-wave long at the center frequency of that band. This antenna is readily tuned for peak signal by means of a series variable capacitor (see *VC* in Fig. 4-4). Depending upon the nature of the input circuit of the receiver, the maximum capacitance of this capacitor will be between 50 and 500 pF.

The inverted-L antenna is uncomplicated and easy to install, and is especially useful when the receiver must be located at one end of an antenna. For further discussion of this type of antenna, see Chapter 2.

T-ANTENNA

When the receiver must be located under the center of a horizontal antenna, the flat top may be operated with a lead-in dropping vertically from the exact center. This results in the T-antenna (Fig. 4-5A), so called from its resemblance to a capital letter *T*. Some technicians regard the T-antenna as two inverted-L antennas in parallel. Thus, in Fig. 4-5B, the left half of the antenna, *abc*, forms the first inverted-L portion; and the second half of the antenna, *abd*, forms the second inverted-L portion. The lead-in, *ab*, is common to both *L*s and may be regarded as composed of separate lead-ins in parallel.

The T-antenna has many features in common with the inverted-L antenna. The flat top should pick up the signal, and the lead-in should conduct signal current to the receiver. The lead-in should pick up as little signal as possible and must be low resistance. This means that the lead-in should be as short as practicable and must run straightaway from the antenna. Unavoidably, however, the lead-in will pick up some signal energy. It is customary to employ bare wire (AWG #12 or #14) for the flat top and insulated wire for the lead-in. A good ground is required for the T-antenna.

Most SWLs who use the T-antenna make the flat top as long as the available space will allow. It is important, however, that the lead-in always be attached to the exact center of the flat top. When most of the listening is done in one band of frequencies, the dimensions *abc* and *abd* in Fig. 4-5B should be a quarter-wavelength at the center frequency of that band.

Like the inverted-L type, the T-antenna may readily be tuned for maximum signal by means of a series variable capacitor (see *VC* in Fig. 4-5A). Depending upon the input circuit of the receiver, the maximum capacitance of *VC* will be between 50 and 500 pF. Again, like the inverted-L type, the T-antenna is uncomplicated and easy to install, and it is especially useful when the receiver must be located under the center of the antenna. For further discussion of the T-antenna, see Chapter 2.

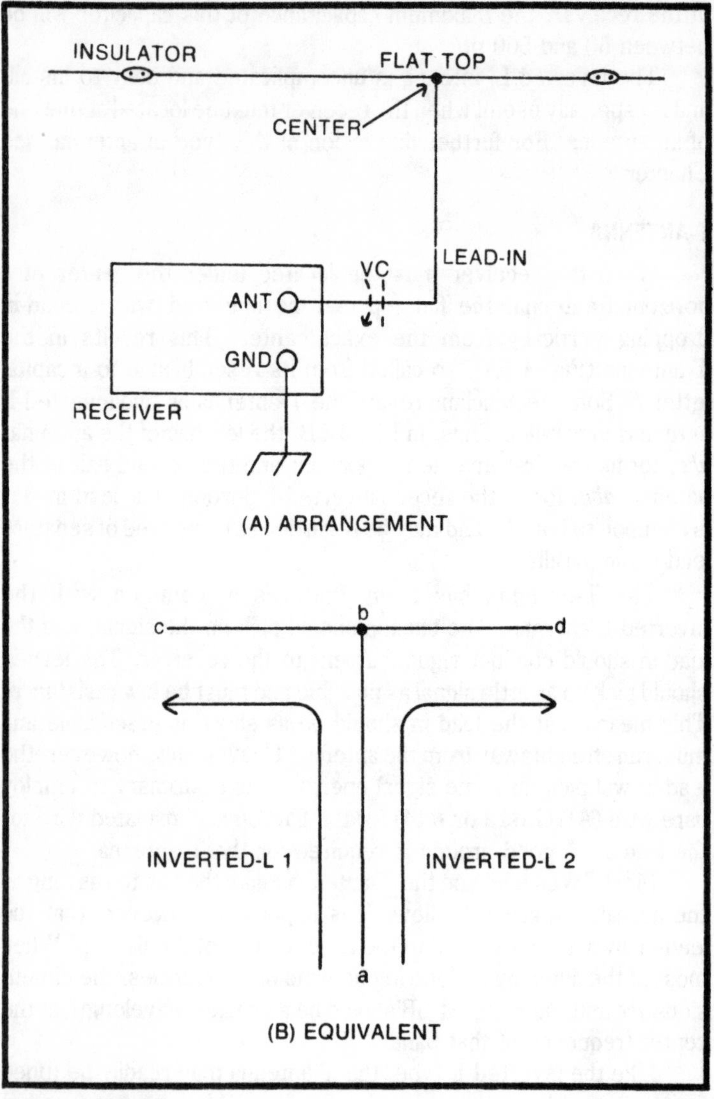

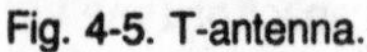
Fig. 4-5. T-antenna.

CROSS ANTENNA

Vertical antennas are not always practicable for SWL use, since for the lower frequencies they need to be too high for many locations. But the vertical antenna is prized for its omnidirectional operation. Omnidirectionality can be approximated with a horizontal installation by connecting two identical-length perpendicular flat tops to a single lead-in to give a sort of double-T antenna, as shown in Fig. 4-6A. As the length of the separate flat tops (north-south and east-west) approaches a half-wavelength, each flat top tends to exhibit two side lobes, and the four resulting lobes combine in a four-leaf-clover pattern, somewhat as in Fig. 4-6B.

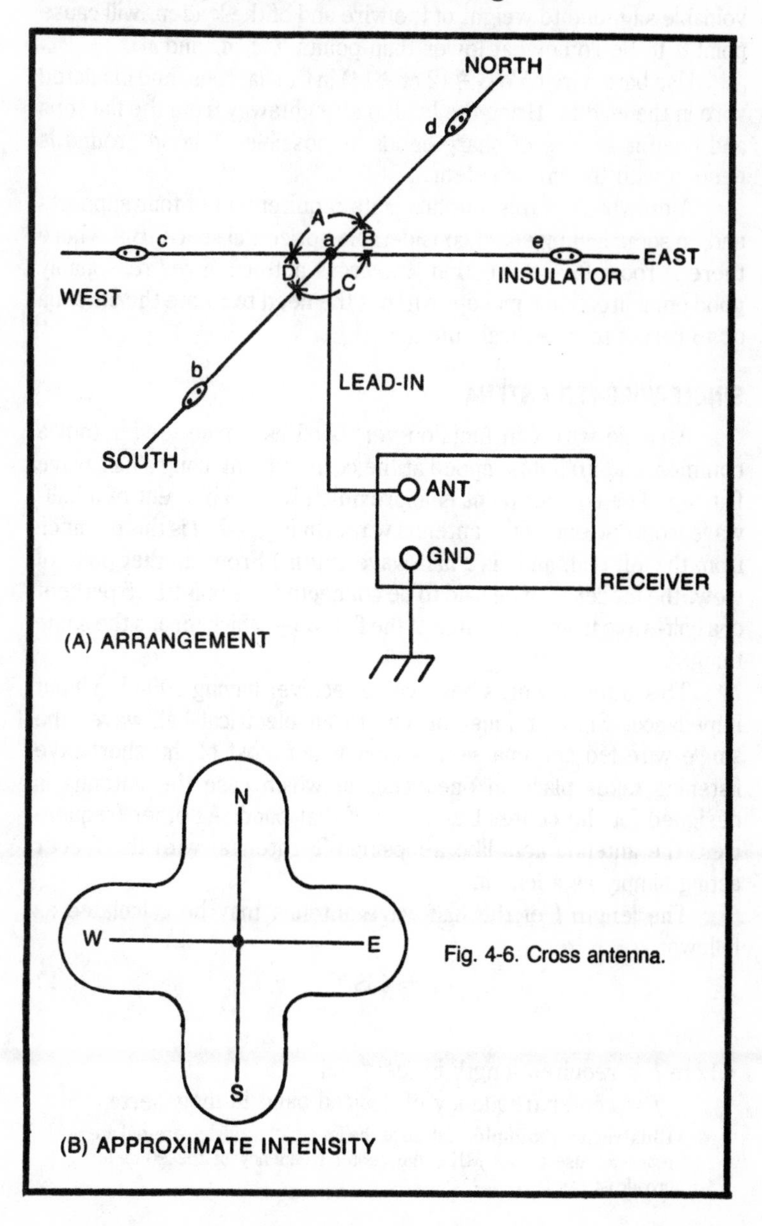

Fig. 4-6. Cross antenna.

In this arrangement, the flat-top lengths are equal; that is, in Fig. 4-6A, $ab = ac = ad = ae$. Also, the flat tops must be exactly at right angles to each other: $\angle A = \angle B = \angle C = \angle D = 90°$. The two flat tops must be connected solidly to the lead-in at point *a*, and each flat top (*bd* and *ce*) should be exactly horizontal. Actually, unavoidable sag, due to weight of the wire and of the lead-in, will cause point *a* to be somewhat lower than points *b*, *c*, *d*, and *e*.)

Use bare wire (AWG #12 or #14) in the flat tops, and insulated wire in the lead-in. Bring the lead-in straightaway from the flat tops and keep it as free of sharp bends as possible. A good ground is needed with the cross antenna.

A drawback of this antenna is its requirement of four supports and, in some instances, of considerable horizontal space. But where there is room for its erection, the cross antenna gives reasonably good omnidirectional pickup, without the need to rotate the antenna or to resort to a vertical antenna.

SINGLE-WIRE-FED ANTENA

A single wire can function very well as a true feeder (not a common lead-in) if it is tapped at the correct point along a half-wave flat top. The correct point is approximately 34.5 percent of a half-wave from one end of the antenna wire. (In Fig. 4-7, *d* is this distance from the left end, and *l* is a half-wavelength.) From another point of view, the feeder may be said to be connected at a point 15.5 percent of a half-wave from the center of the flat top—which means the same thing.

This antenna works best with a receiver having 300-ohm input impedance. Since it must be cut to an electrical half-wave, the single-wire-fed antenna serves best when most of the shortwave listening takes place in one band, in which case the antenna is designed for the center frequency of that band. At other frequencies, the antenna acts like a lopsided T-antenna, with the feeder acting simply as a lead-in.

The length *l* of the half-wave antenna may be calculated as follows:

$$l = 468/f \qquad (4\text{-}2)$$

where l = required length in feet, and
f = center frequency of desired band in megahertz.

Illustrative Example. Calculate the length of a single-wire-fed antenna for use at 3.3 MHz, the center frequency of the 90-meter broadcast band.

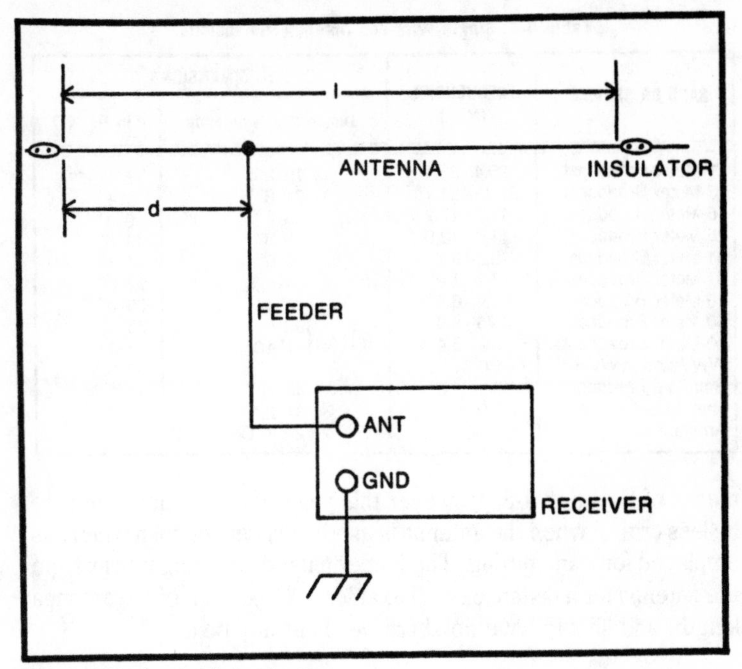

Fig. 4-7. Single-wire-fed antenna.

From Equation 4-2,

$$f = 468/3.3 = 141.82 \text{ feet} = 141 \text{ feet } 10 \text{ inches.}$$

The distance from one end of the antenna to the point at which the feeder must be connected may be calculated as follows:

$$d = 161.5/f \tag{4-3}$$

where d = distance from one end of antenna wire, in feet, and
f = center frequency of desired band in megahertz.

Illustrative Example. Calculate the distance between the end of the above antenna and the point of connection of the single-wire feeder.

From Equation 4-3,

$$d = 161.5/3.3 = 48.84 \text{ feet} = 48 \text{ feet } 11 \text{ inches.}$$

For convenience, column 3 of Table 4-1 gives the length of a half-wave antenna for the center frequency in nine shortwave broadcast bands and at the three frequencies employed by Standard Frequency Stations WWV and WWVH. Column 4 of the same table gives the distance from one end of the antenna to the point at which the single-wire feeder must be connected.

At some locations, the *exact* point of connection of the feeder must be located experimentally after d has been determined by

Table 4-1. Single-Wire-Fed Antenna Dimensions.

BAND OR SERVICE	FREQUENCIES (MHz)	(l) DIMENSIONS	
		Length of λ/2 Antenna	d in Fig. 4-7
11-Meter Broadcast	25.4–26.1	18′ 2″	6′ 3.25″
13-Meter Broadcast	21.45–21.75	21′ 8″	7′ 6″
16-Meter Broadcast	17.7–17.9	26′ 3.5″	9′ 1″
25-Meter Broadcast	11.6–12.0	39′ 8″	13′ 8″
31-Meter Broadcast	9.2–9.7	49′ 6″	17′ 1″
41-Meter Broadcast	7.1–7.4	64′ 6.5″	22′ 3″
49-Meter Broadcast	5.9–6.4	76′ 1″	26′ 3″
60-Meter Broadcast	4.75–5.0	96′ 0″	33′ 1″
90-Meter Broadcast	3.2–3.4	141′ 10″	48′ 11″
WWV and WWVH	5.0	93′ 7″	32′ 3.5″
Standard Frequency,	10.0	46′ 10″	16′ 1.75″
time, etc.	15.0	31′ 2.5″	10′ 9″
Amateur	See Fig. 2-12B in Chapter 2.		

means of Equation 4-3. However this point of connection seems to be less critical when the antenna is used for receiving than when it is employed for transmitting. The feeder must drop straightaway from the antenna for a distance equal to at least 44 percent of the antenna length, and should have no sharp bends at any point.

STANDARD DIPOLE ANTENNA

When principal SWL activity is confined to a single band and compromise operation at other frequencies is acceptable, a *dipole* antenna designed for the principal band is simple, efficient, and easy to work with. This is a horizontal, half-wave antenna split at its center and connected there to a coaxial line (any length). See Fig. 4-8.

A 75-ohm coaxial line, such as type RG59, is employed. This line may be any convenient length and it matches the center of the antenna and also the 75-ohm input of the receiver. The length l of each half of the antenna may be calculated as follows:

$$l = 234/f \qquad (4\text{-}4)$$

where l = required length in feet, and

f = center frequency of the principal band in megahertz.

Illustrative Example. It is desired to monitor the frequencies between 8 and 17 MHz. Calcualte the length which each half of a dipole antenna must have for this service.

(a) The center frequency of this range is 12.5 MHz. This is the value f to be used in the calculation.

(b) From Equation 4-4,

$$l = 234/12.5 = 18.72 \text{ feet} = 18 \text{ feet } 8.5 \text{ inches}$$

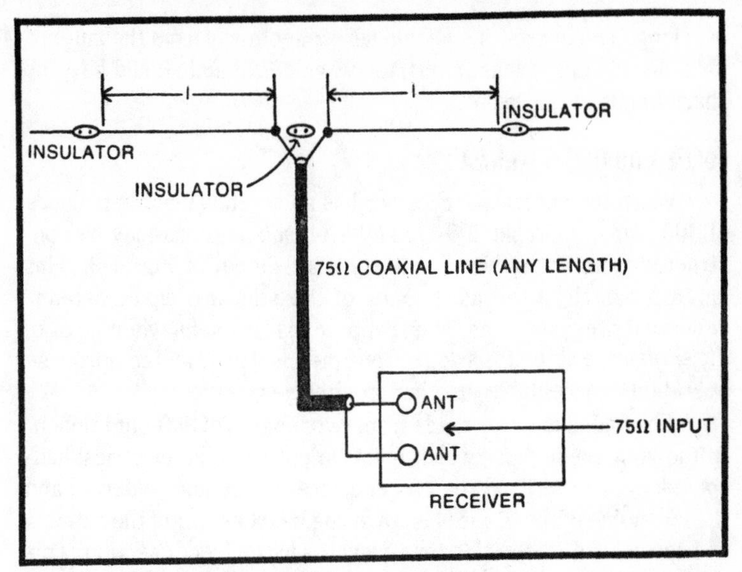

Fig. 4-8. Standard dipole antenna.

Column 3 in Table 4-2 gives length *l* for the center frequency in nine shortwave broadcast bands and for the frequencies of WWV and WWVH standard frequency and standard time broadcasts.

The dipole antenna should be made with AWG #12 or #14 wire. The three strain insulators used with this antenna are each 1.5

Table 4-2. Dipole Antenna Dimensions.

BAND OR SERVICE	FREQUENCIES	LENGTH (l in Fig. 4-8 and 4-9)	
		CONVENTIONAL DIPOLE FIG. 4-8	FOLDED DIPOLE FIG. 4-9
11-Meter Broadcast	25.4–26.1 MHz	9′ 1″	7′ 10″
13-Meter Broadcast	21.45–21.75	10′ 10″	9′ 4″
16-Meter Broadcast	17.7–17.9	13′ 1.75″	11′ 4″
25-Meter Broadcast	11.6–12.0	19′ 10″	17′ 1″
31-Meter Broadcast	9.2–9.7	24′ 9″	21′ 4″
41-Meter Broadcast	7.1–7.4	32′ 3″	27′ 10″
49-Meter Broadcast	5.9–6.4	38′ 6″	32′ 9.5″
60-Meter Broadcast	4.75–5.0	48′ 0″	41′ 4.5″
90-Meter Broadcast	3.2–3.4	70′ 11″	61′ 1″
WWV and WWVH	5.0	46′ 9.5″	40′ 4″
Standard-Frequency,	10.0	23′ 4.75″	20′ 2″
Time, Etc.	15.0	15′ 7.25″	13′ 5″
Amateur Bands	For conventional dipole, see Fig. 2-5 (C) in Chapter 2. For folded dipole, see Fig. 2-6(B) in Chapter 2.		

inch long. The coaxial line should fall straightaway from the antenna for a distance of at least a quarter-wavelength and should have no sharp bends at any point.

FOLDED DIPOLE ANTENNA

When the shortwave receiver has an antenna input impedance of 300 ohms, a simple 300-ohm *folded dipole* antenna may be constructed from flat-ribbon TV lead-in, as shown in Fig. 4-9. This antenna has all of the advantages of the standard dipole antenna described previously and, like that antenna, is useful when most of the shortwave listening is done in one principal band and compromise operation is acceptable at other frequencies.

The folded dipole is made from two lengths of 300-ohm ribbon. In the antenna section, which must be cut to be an electrical half-wave long, the wires at the two ends are spliced and soldered, and the lower one of these wires is cut at the exact center of the antenna and spliced and soldered to the two wires in the feeder section. This construction is shown in Fig. 4-9.

The length l of each half of the antenna may be calculated as follows:

$$l = 201.7/f \qquad (4\text{-}5)$$

where l = required length in feet, and
f = center frequency of principal band in megahertz.

Illustrative Example. It is desired to monitor frequencies between 5 and 7 MHz. Calculate the length which each half of a folded dipole antenna must have for this service.

(a) The center frequency of this range is 6 MHz. This is the value f to be used in the calculation.

(b) From Equation 4-5,

$$l = 201.7/6 = 33.62 \text{ feet} = 33 \text{ feet } 7.5 \text{ inches.}$$

Column 4 in Table 4-2 gives length l for the center frequency in nine shortwave broadcast bands and for the frequencies of WWV and WWVH standard frequency and standard time broadcasts.

The feeder section of the ribbon should fall straightaway from the antenna section for a distance of at least quarter-wavelength and should have no sharp bends at any point.

COMMON VERTICAL ANTENNA

The common vertical antenna consists simply of a single vertically stretched wire or a rod insulated from ground and surrounding structures, with a lead-in to the receiver at its lower end. This

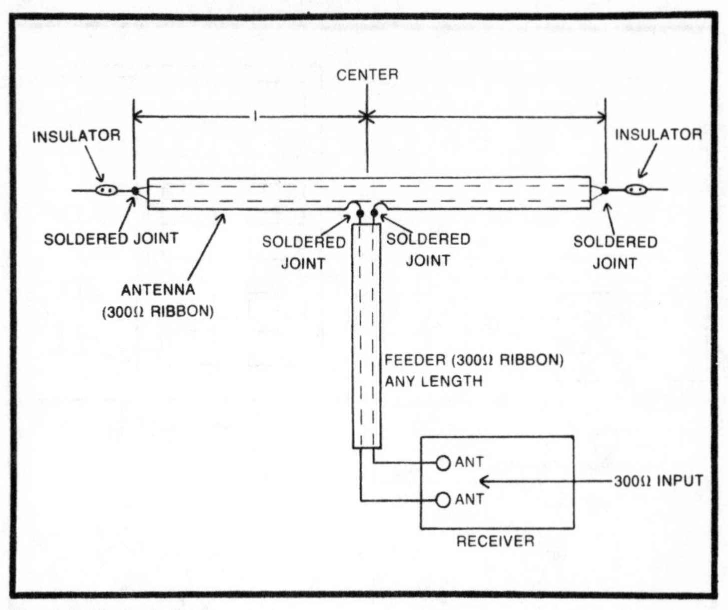

Fig. 4-9. Folded dipole.

antenna (see Fig. 4-10A) resembles an inverted-L antenna standing on one end.

For general-purpose SWL use, the antenna should be as long as possible and the lead-in as short as possible. A good ground is imperative. This simple vertical antenna is omnidirectional, but it is sensitive to electrical noise and must be situated as far as possible from power leaks and interfering machinery. Moreover, it is a veritable lightning rod and should always be equipped wth a good lightning arrester or lightning switch.

Where horizontal space is at a premium, the vertical is a life saver antenna. Figures 4-10B, 4-10C, and 4-10D show a few of the ways of erecting a common vertical antenna. An often-used indoor version consists simply of an insulated wire tacked or taped down a wall of the radio room.

GROUND-PLANE ANTENNA

The *ground-plane* antenna, so familiar in citizens band work, performs exellently in a single shortwave band and gives compromise operation elsewhere. It consists of a quarter-wave "radiator" (the term is borrowed from use of the antenna for transmitting) at the base of which are spaced several (usually 3 or 4) horizontal radials (wires or rods which are connected to each other

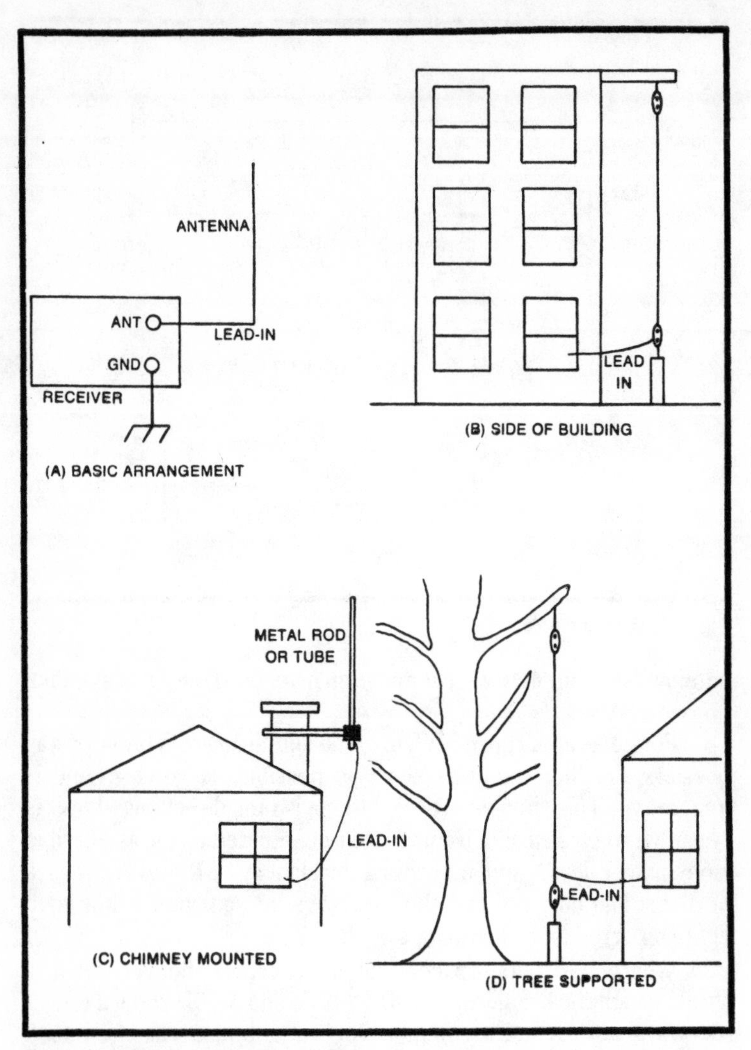

Fig. 4-10. Common vertical antenna.

and grounded, but insulated from the radiator. This assembly of radials constitutes the ground plane, which in effect brings ground up to the base of the antenna.

Figure 4-11A shows this arrangement. The radiator is connected to the center conductor of a coaxial line of any length, and the radials are connected to the outer metal sheath of the line. The sheath is grounded at its lower end. If the radials are bent downward at an angle of about 45°, a 52-ohm line—such as type RG8/A-AU—can be used without any impedance-matching device.

In this antenna, the radiator is a quarter-wavelength long at the center frequency of the band of interest, and each ground-plane radial is 4 percent to 6 percent longer than the radiator. The ground-plane antenna is limited to those frequencies at which these lengths will be practical at a given location. Figure 4-11B gives dimensions for radiator and radials at the center frequency of the 11-, 13-, 16-, 25-, and 31-meter broadcast bands. At lower frequencies (longer wavelengths) than these, the lengths start to grow unwieldy.

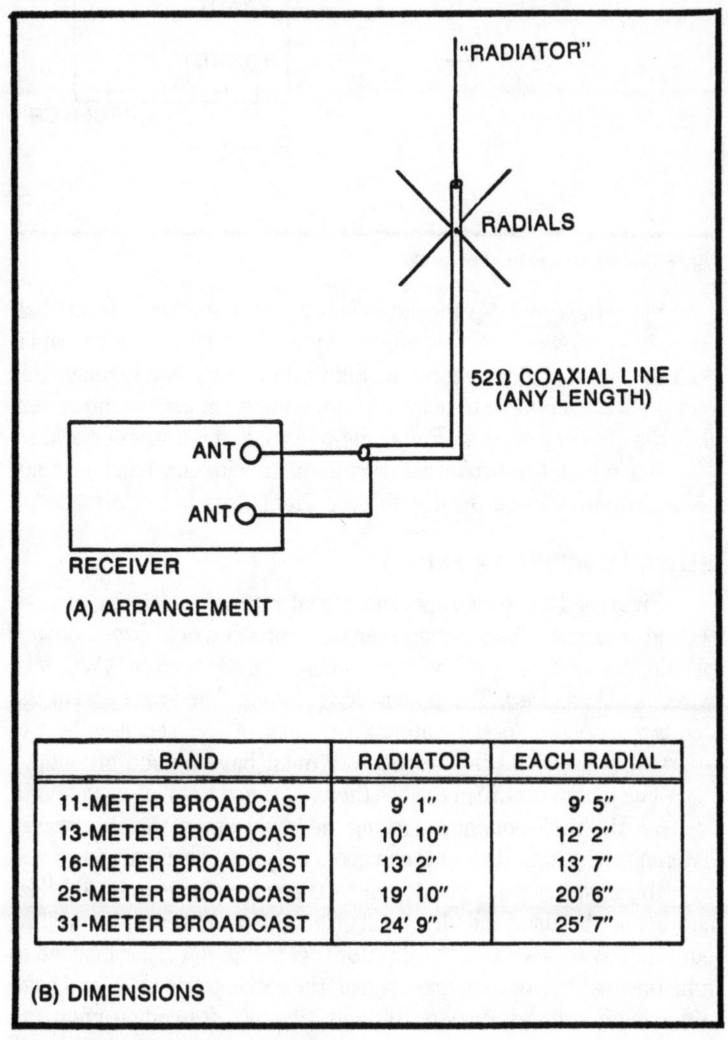

BAND	RADIATOR	EACH RADIAL
11-METER BROADCAST	9' 1"	9' 5"
13-METER BROADCAST	10' 10"	12' 2"
16-METER BROADCAST	13' 2"	13' 7"
25-METER BROADCAST	19' 10"	20' 6"
31-METER BROADCAST	24' 9"	25' 7"

(B) DIMENSIONS

Fig. 4-11. Ground-plane antenna.

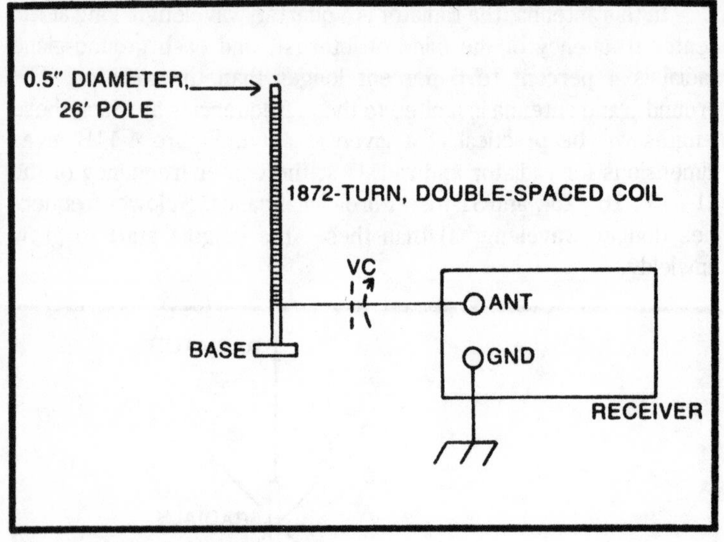

Fig. 4-12. Helically wound antenna.

Since the ground-plane antenna is vertical, it is omnidirectional. Also, it is endowed with low-angle response, and this feature suits it well to long-distance reception. Like other vertical antennas, this one is also susceptible to noise pickup, somewhat less so, however, than the ordinary vertical. For single-band use, it is a star performer.

For 11-meter broadcast reception, a citizens band ground-plane antenna can be used with only slight drop in performance.

HELICALLY WOUND ANTENNA

Figure 4-12 shows a special vertical antenna consisting of a coil wound around a half-inch-diameter waterproofed pole (wood dowel, plastic, bamboo, etc.). The coil contains 1872 turns of AWG #12 wire, double spaced. The pole is 26 feet long. The lower end of the coil is connected to the antenna terminal of the receiver by the shortest possible lead. The receiver must have a good ground.

The length of wire in the coil (helix) is equal to a half-wavelength at 1.6 MHz, the frequency marking the lower end of the shortwave-listening spectrum. This antenna gives excellent performance at and near this frequency and good performance elsewhere up to 30 MHz, and it may be used indoors (ceiling height permitting), as well as outdoors. A series variable capacitor (*VC* in Fig. 4-12) can be used to tune the antenna for maximum signal; the capacitance of *VC* will have a maximum value between 100 and 400 pF, depending upon the inductance of the antenna coil in the receiver.

The helically wound antenna is less susceptible to pickup of electrical noise than is the conventional vertical antenna.

WEATHER ANTENNAS

Shortwave listeners who principally monitor the broadcasts and alarms of the U. S. National Weather Service in the 162.4–162.55 MHz band will find the two antennas in Fig. 4-13 invaluable for their use. Each is designed for the center frequency (162.475 MHz) of the weather band.

Figure 4-13A shows a standard dipole consisting of two straight, rigid, 1-foot 5.25-inch sections of AWG #10 copper wire. The near ends of these sections are held by a 3 × 5-inch plastic plate which is also provided with a bracket to hold the assembly to a mast. On the plate, the antenna sections are separated by a half-inch, and a 73-ohm coaxial line (for example, type RG59) is attached to the sections. This coax may be any desired length. This antenna matches a receiver having 75-ohm input impedance closely enough for all practical purposes. Because of the small size of the antenna (2 feet 11 inches overall), it may easily be rotated for maximum signal.

Figure 4-13B shows a folded dipole consisting of two lengths of 300-ohm TV ribbon, one length serving as the antenna, the other as the feeder. The antenna section is 2 feet 6 inches long. The ends of the two wires in this section are soldered together, as shown at *X* and *Y*. The lower wire in the antenna section is cut at its exact center, and the two resulting wires are soldered to the two wires in the feeder, as shown at each point *Z*. This antenna matches a receiver having an input impedance of 300 ohms.

MISCELLANEOUS ANTENNAS

Various other antennas—some permanent, some temporary; some makeshift, and some regular—are used with shortwave receivers. Some of these antennas are described in the following paragraphs.

Molding Antenna

This inside antenna consists of an insulated flexible wire taped or tacked around the molding of a room. Best results usually are obtained in an upstairs room, for then the antenna is farthest from ground. At some locations, a single side of the room is preferable, while at others the wire must be run around all four sides for maximum signal. Experiment—first with each side of the room, then

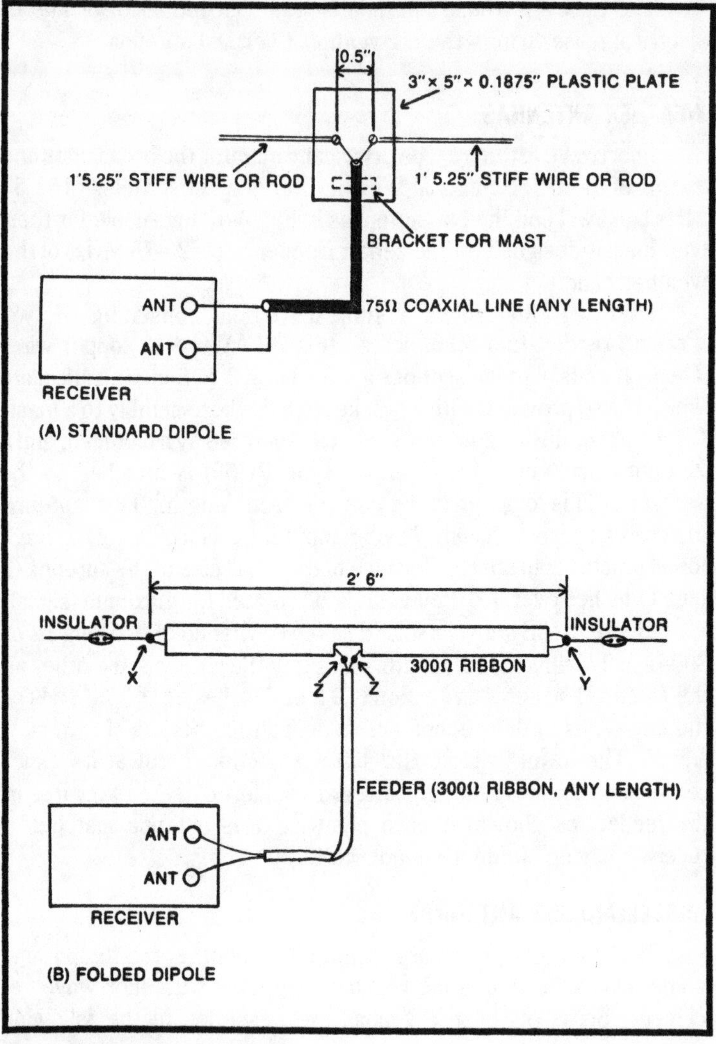

Fig. 4-13. Weather antennas.

with all four sides. Connect one end of this antenna to the ANTENNA terminal of the receiver. Use a good ground.

Door-Frame Antenna

This inside antenna consists of a flexible insulated wire taped or tacked around the frame of a door. Best results usually are obtained in an upstairs room because then the antenna is farthest from ground. At some locations, best results are obtained by using only one of the vertical sides of the door; at others, one side and top; and at still others, both sides and the top. Experiment for maximum signal. Connect one end of this antenna to the antenna terminal of the receiver. Use a good ground.

Window-Frame Antenna

This inside antenna consists of a flexible insulated wire taped or tacked around the *nonmetal* frame of a window. Best results usually are obtained in an upstairs room because then the antenna is farthest from ground. Experiment to find which of the following arrangements gives maximum signal: wire running on one vertical side of the window, one vertical side and the horizontal top, one vertical side and the sill, both vertical sides and the top, or all four sides. Connect one end of the wire to the antenna terminal of the receiver. Use a good ground.

Telescoping Whip

A small, telescoping whip antenna, such as shown in Fig. 4-14A may be bought as a separate component that extends to 36 inches and collapses down to 6 or 7 inches. This vertical antenna—often stood on top of the receiver by means of a suction cup—is connected directly to the antenna terminal of the receiver. While this type of antenna is classically an indoor type, it sometimes is placed outside of a window. The whip is a short antenna and therefore gives compromise operation at most points in the shortwave-listening spectrum; when extended to its full 36 inches, for example, it provides half-wave operation at the high frequency of 156 MHz. A good ground must be used with this antenna.

Built-in Whip

Some short-wave receivers, such as Heathkit GR-78, have a built-in telescoping whip, as shown in Fig. 4-14B. This antenna projects from the top of the case, usually extending to a maximum length of about 36 inches, and often collapses fully into the case. The remarks concerning the external whip in the frequency discussion apply equally well to the built-in whip. Use a good ground with this antenna, unless the receiver is being carried about.

Rabbit-Ears Antenna

This two-element telescoping antenna (see Fig. 4-14C) is familiar as an accessory for TV and FM receivers. It consists of two telescoping legs mounted on a hinged joint which permits the angle of

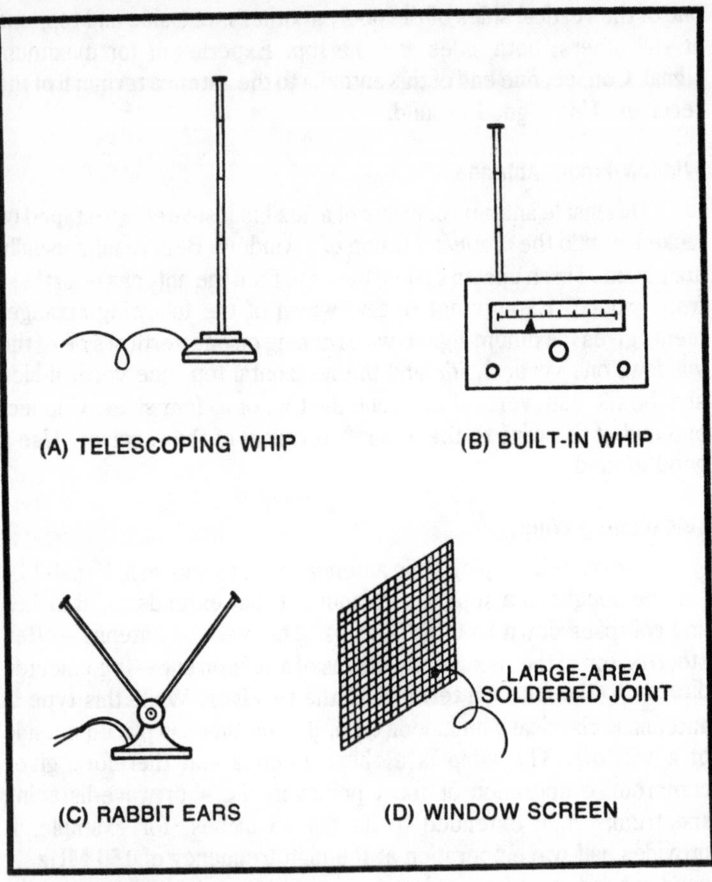

Fig. 4-14. Miscellaneous antennas.

separation between them to be varied smoothly. This antenna may also be rotated. It is a dipole with a short length of cable connected to the base of the legs (the two leads at the other end of the cable are connected to the antenna terminals of the receiver). This antenna may be set on top of the receiver or on a nearby table.

Rabbit ears offer considerable flexibility, since the length of each leg can be adjusted independently of the other (usually up to 35 or 45 inches, depending upon make and model); the angle between the legs is adjustable (almost from 0° to 180°); and the antenna is rotatable. The operator should work with all of these features, searching for maximum signal.

Like the telescoping whip antenna, however, the rabbit-ears antenna is a *short* antenna at most points in the shortwave-listening spectrum, hence gives compromise operation. When the antenna is extended to its full 45 inches per leg, it provides half-wave operation at the high frequency of 62.4 MHz.

Window-Screen Antenna

A tightly woven metal window screen which is well insulated from ground makes a fairly good emergency antenna (see Fig. 4-14D). Solder a short insulated wire to the screen and connect it to the antenna terminal of the receiver.

This antenna usually works best at an upstairs window since it then is highest above ground and more in the clear than when it is downstairs. Also, a screen having unpainted wire works better than the painted variety. A good ground is required with this antenna.

Random Wire

Twenty feet of thin, flexible, insulated wire may be wound around a card and unreeled indoors or outdoors at will for a temporary antenna of surprising merit. With the end of the hank connected to the antenna terminal of the receiver, the card end is hung or tacked across the room to a door, window wall, or ceiling (or outdoors to a building, tree, fence, or the like), as preferred. The receiver must have a good ground, for best results.

This antenna can be run at any angle (vertical or horizontal) that will give maximum signal, and can quickly be disconnected, rolled up, and tucked away when not in use.

Power Line Prohibited

Some SWLs have had moderate success using the power line as an antenna. They connect the ungrounded side of the power line to the antenna terminal of the receiver through a high-voltage series capacitor, usually 0.002 to 0.005 μF. The reactance corresponding to this capacitance is high enough at 60 Hz that any current through the antenna coil will be of the order of microamperes, but it is low enough at radio frequencies that shortwave signals (if any are picked up by the power line) pass through the capacitor readily.

Sometimes, this scheme works, and often it does not. *It is not an entirely safe practice, however, and this book does not recommend it.* If the capacitor does happen to leak, the shortwave listener is exposed to dangerous electric shock, and the receiver may be damaged.

TUNING THE ANTENNA

The simple expedient of tuning the antenna can often produce a large increase in signal strength. This is especially true when a

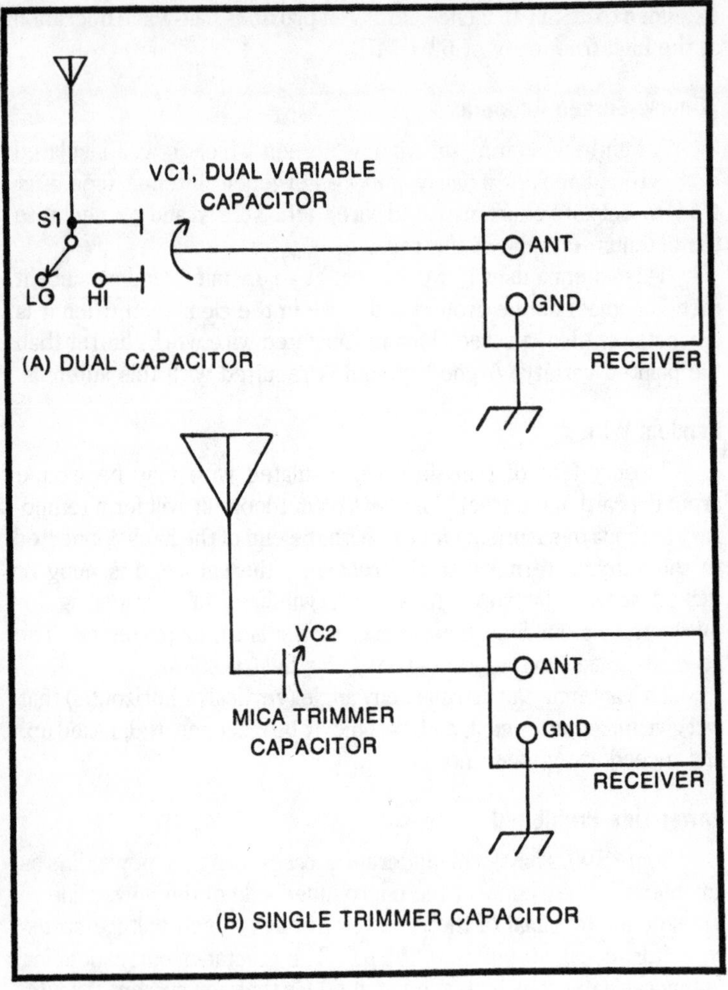

Fig. 4-15. Tuning the antenna.

compromise antenna must be used. When the receiver contains an antenna coil, tuning can be accomplished by means of a variable capacitor connected in series with the antenna and receiver, as shown in Fig. 4-15. This scheme allows the entire input circuit (antenna plus capacitor plus antenna coil) to be adjusted to series resonance at the signal frequency; at series resonance, maximum signal current flows through the antenna coil.

The required tuning capacitance is governed by the inductance of the antenna coil in the receiver; and since this inductance is not usually specified in receiver literature, it cannot easily be known by most SWLs. Hence, a tuning capacitor with high maximum capacitance is recommended to accommodate the wide variety of antenna coils.

In Fig. 4-15A, the dual variable capacitor, VC1, has a maximum capacitance of 432 pF per section and a minimum capacitance of 12.5 pF per section. When switch S1 is in its *LO* position, a single 12.5–432 pF section of this capacitor is used; when S1 is in its *HI* position, both sections are automatically connected in parallel and the capacitor has a range of 25–864 pF. (A suitable variable capacitor with these specifications is Allied Electronics model S3.)

In Fig. 4-15B, a small screwdriver-adjusted mica trimmer capacitor (VC2) is employed. This unit provides a maximum capacitance of 200 pF and a minimum capacitance of 24 pF. (A suitable trimmer wth these specifications is Allied Electronics model 425.)

With variable-capacitor tuning, the procedure is simple: (1) With the tuning capacitor randomly set, tune in the desired signal, using the receiver controls only. (2) Then, without touching the receiver setting, adjust the capacitor for maximum signal.

Chapter 5
Construction Notes

Each reader brings to the building and erection of an antenna his own mechanical skills and preferences. The requirements of sturdiness, safety, and eye appeal can be met in a number of ways, and in some instances no way can be said to be better than the rest. In short, there often is no one best way to build an antenna. It is chiefly for the benefit of the newcomer and the nonspecialist, therefore, that this chapter offers construction hints. Every attempt has been made to keep this material relevant.

In addition to the items in this chapter, occasional short construction hints for specific antennas may be found in earlier chapters.

CONDUCTOR SIZE

For electrical efficiency and mechanical strength, it is essential that conductors of the proper size and type be used in antennas. These conductors consist of wire, tubing, and sometimes rod.

Wire

In general, thick wire is required for a transmitting antenna because the transmitting antenna must carry appreciable current; thin wire may be used for a receiving antenna because the signal current in this antenna is very low. But this generalization is subject to modification. For example, thin wire is not satisfactory for an *outside* receiving antenna that might snap under the load of ice, but might be entirely suitable for an inside antenna.

AWG #12 and #14 (and occasionally #16) hard-drawn copper wire is commonly used for outside antennas, especially the transmitting type. The thinner AWG #18 and #20 are useful for receiving antennas, especially the inside variety. Although stranded wire is easier to pull straight than is solid wire, it is less desirable since broken strands change the RF resistance of the wire and can also introduce receiver noise. For added mechanical strength, steel-core copper (trade name *Copperweld*) wire is best for outside antennas (the DC resistance of the steel core is unimportant, as the RF currents flow in the copper skin). In an open-wire transmission line, the exact specified size of wire must be used, since the impedance of the line depends upon the wire diameter (see, for example, Equation 1-4 in Chapter 1). Customarily, bare wire is used in the flat top and plastic-covered wire of the same size in the conventional lead-in. Unlike the lead-in, however, 1- or 2-wire feeders are made of the same size bare wire as the radiator or flat top.

Regarding the reactive effect of a straight wire, such as is employed in an antenna or feeder, a straight 100-foot length of AWG #12 wire has an inductance of approximately 62 μH; if this wire is mounted horizontally 40 feet above ground, it exhibits a capacitance of approximately 1611 pF. By comparison, the same length of AWG #20 wire 40 feet above ground exhibits 68 μH and 1315 pF. Table 5-1 gives important characteristics of AWG #12, #14, #16, #18, and #20 bare copper wire.

Tubing

As a rule, vertical antennas and small horizontal rotary antennas are built with metal tubing, usually aluminum, stainless steel, or copper. Common outside diameters of this tubing range from 0.1875 to 1.25 inches, the choice depending upon length of elements (in long elements, for example, the larger diameters result in less sag), desired overall size, and desired overall weight. Common wall thicknesses are 0.028, 0.035, 0.049, 0.058, 0.065, and 0.083 inch, but sometimes 0.125 and 0.250 inch also are used. When a telescoping antenna element is desired, if all sections of the tubing are of 0.058-inch wall thickness, each section will telescope well into the next larger diameter section. Sometimes, tubing and wire are combined in an antenna. Very small directional antennas often employ short pieces of AWG #10 copper wire (OD = 101.9 mils), AWG #12 copper wire (OD = 80.8 mils), eighth-inch-diameter welding rod or hard-drawn aluminum wire.

Table 5-1. Copper Wire Characteristics.

AWG	DIAMETER (mils)	MAXIMUM AMPERES	FEET PER POUND	OHMS PER FOOT (25°C)
12	80.81	41	50.59	0.0016
14	64.01	32	80.44	0.0025
16	50.80	22	127.9	0.0041
18	40.30	16	203.4	0.0064
20	31.96	11	323.4	0.0102

INSULATION AND INSULATORS

Any antenna—transmitting or receiving—must be well insulated from ground and nearby objects, such as masts or towers. Antenna insulators must be weatherproof, and break resistant.

Figure 5-1 shows a number of insulators used in antenna installations. The *strain* insulator (Fig. 5-1A) is often found at each end of an antenna wire, the latter being passed through one hole of the insulator and the support wire through the other hole. This type may also be used in guy wires. The chief advantage of the strain insulator is the long surface leakage path provided by the corrugations. Strain insulators are available in glass and in glazed ceramic, and in various sizes. For a very long leakage path, use the longest strain insulator obtainable, or connect two insulators in series, or both. The strain insulator has the disadvantage that when it breaks in two, the two wires fall.

Another insulator that is widely used in antenna wires and guy wires is shown in Fig. 5-1B. This is the ceramic *egg* insulator. One wire is tied around the outer surface of the egg in a groove provided for the purpose, and the other wire is passed through the central eye of the egg. This arrangement has the advantage that if the insulator breaks in two, the two wires interlink with each other and do not fall.

Various insulators are available for supporting a lead-in. Fig. 5-1C shows the *ceramic knob*. This insulator has a top section and bottom section that separate or are held together with a nail or screw that secures the knob to a mounting surface. Lead-in wires pass through the two mating grooves in the top and bottom sections. Pass the wires through the grooves in the bottom section, place the top section in place, and fasten the knob securely to the building with a nail (usually supplied) passed through the center hole of the in-

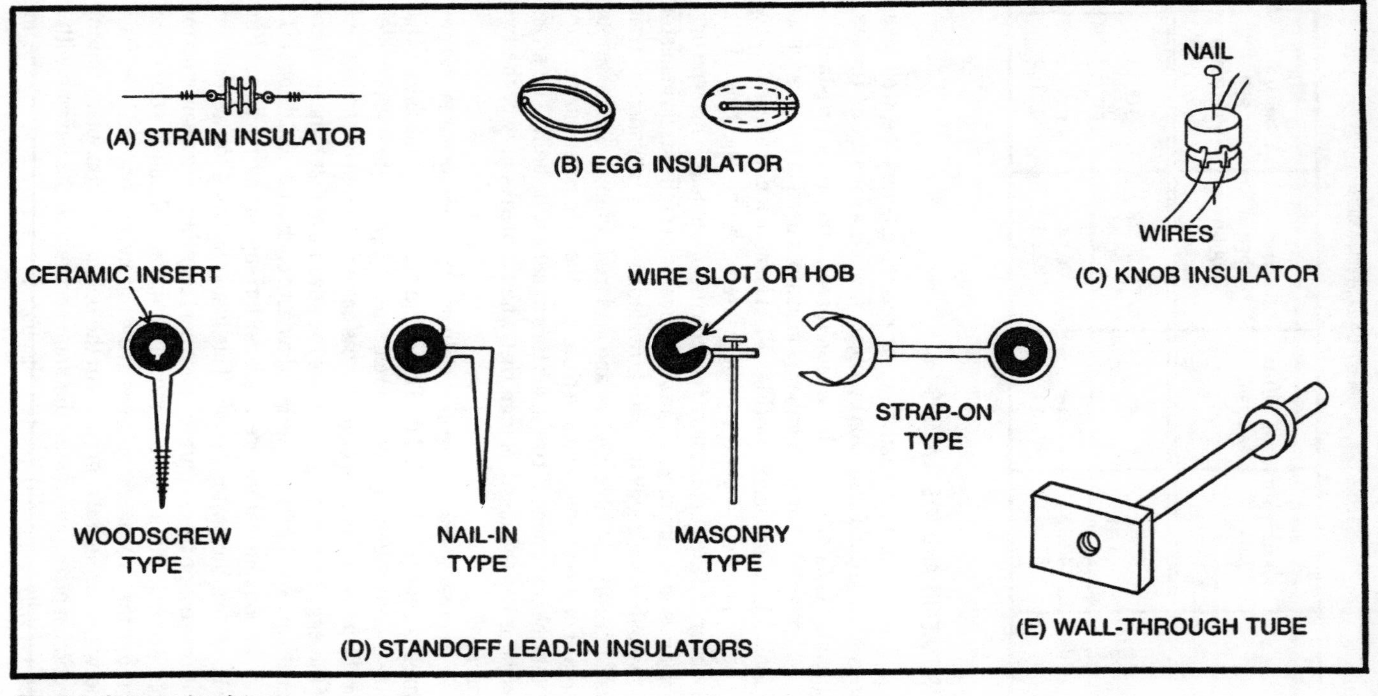

Fig. 5-1. Antenna insulators.

sulator. If glazed and unglazed knobs both are available, use the glazed version for its superior insulating qualities and water resistance. Other lead-in insulators of the *standoff* type are shown in Fig. 5-1D. These consist essentially of a ceramic or plastic insert in a metal bracket. The lead-in wire is passed through a hole or slot in the insert. (The aperture is round for standard wire or slotted for flat ribbon.) The first insulator in Fig. 5-1D terminates in a wood screw for attachment to a frame structure. The second is equipped with a strong, sharp spike for driving into brick or concrete. The third has a nail for simple attachment to a supporting structure. The fourth terminates in a curved spring collar which straps around a pipe or pole. (Some strap-on insulators have a screw for tightening the split ring; others grip the pipe by spring action alone.) All of these components in Fig. 5-1D enable the lead-in to be supported away from the building or the antenna mast and they hold it steady.

Figure 5-1E shows a tube-type insulator for bringing the lead-in through the wall of a building. This type fits wall thicknesses up to 13 inches and is grommet sealed.

CONNECTIONS AND SPLICES

It is best not to splice any part of an antenna wire; instead, use a strand of wire that is long enough in the first place. In some instances, however, as in the connection of a lead-in or transmission line to the center of the antenna, splices are unavoidable, and these must be made with the utmost care.

Figure 5-2 shows the three steps in making a sturdy electrical splice. First, clean the two wires thoroughly and tin them (do not use corrosive flux). Second, wind one wire *tightly* around the other (do not depend upon subsequent soldering to hold them together, but make a solid mechanical joint now). The purpose of the spaces between the turns of wire is to allow solder to flow around and between the turns. Use enough turns to make the joint at least 1 inch long. Third, solder the entire joint, completely encasing all the turns. Do not use corrosive flux.

Transmitters, receivers, relays, lightning protectors, outlets, terminal strips, and similar devices are often equipped with screw-type terminals. This allows rapid, simple connection and disconnection, the wire being held under the head of the tightened screw. The wire must be cleaned thoroughly and the screw tightened firmly to insure a good connection with this type of terminal. Solderless, screw-type terminals are employed in wall plates (see Fig. 5-3). These plates, which may be fastened to a wall or baseboard in a room

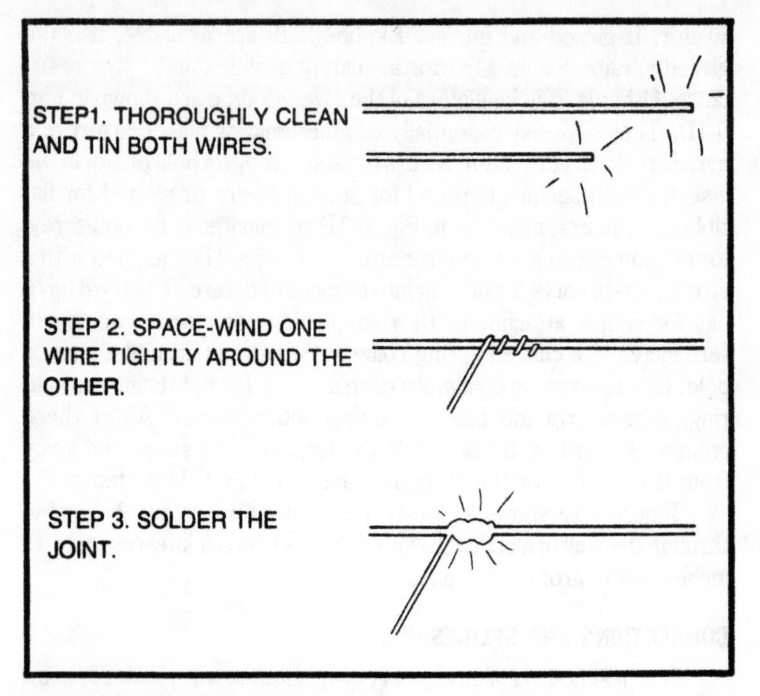

STEP1. THOROUGHLY CLEAN
AND TIN BOTH WIRES.
STEP 2. SPACE-WIND ONE
WIRE TIGHTLY AROUND THE
OTHER.
STEP 3. SOLDER THE
JOINT.

Fig. 5-2. Making an electrical joint.

to allow the antenna line to be brought to a neat access point, promote efficiency as well as good appearence. The line from the transmitter or to the receiver is terminated with a plug which mates with the jack. Figure 5-3A shows a coaxial-type plate, and Fig. 5-3B shows a ribbon-type plate (with mating plug).

MASTS AND OTHER SUPPORTS

Careful attention must be given to the choice and erection of structures which will support an antenna. There are many to choose from, but the ideal support is the one that is best for a given location.

Masts

The simplest mast consists of a pole, pipe, or timber of the desired length, set into the ground or held at its base by a suitable attachment. Except when the mast is comparatively short and is strong enough to remain vertical against the weight and pull of the antenna and the force of winds, it must be well guyed.

Easily available materials are used for simple masts. These are metal pipes, bamboo poles, wood dowels, carpet poles, and timbers.

For increased length, separate members may easily be fastened together. Thus, in Fig. 5-4A, a 40-foot mast is made by splicing together two 20-foot 2 × 3s. Two 2-foot pieces of 2 × 3 are used, as shown, as the fastening strips, being held to the butted mast sections by four carriage bolts. Two lengths of pipe may be similarly joined by means of a threaded coupling, as shown in Fig. 5-4B. Some builders advocate separately guying spliced masts at the point of splice.

Figure 5-4C shows a simple wooden tilt-over mast that is popular with radio hams. The mast rests between two 2-foot 2 × 3s and is held by two carriage bolts. When one of these bolts is temporarily removed, the mast may be swung down, the second bolt acting as a pivot, for work on the antenna.

Commercially available telescoping masts may be obtained in several maximum lengths up to 36 feet. These masts are made of interlocking sections of galvanized steel and have rings at strategic points for attaching guy wires. Also available commercially is steel masting which may be obtained in 5-foot and 10-foot sections which can be stacked, and 5-foot masts of polyester-finished 19-gauge aluminum.

Metal masts must be grounded and must be painted for protection against the elements. Wooden masts also must be painted for weatherproofing.

For both mechanical and aesthetic reasons, a mast must be straight. Don't guess about it—use a plumb line to insure that the mast is vertical.

For rotary antennas, select and erect a mast that is sturdy enough to support the rotator of your choice and the antenna;

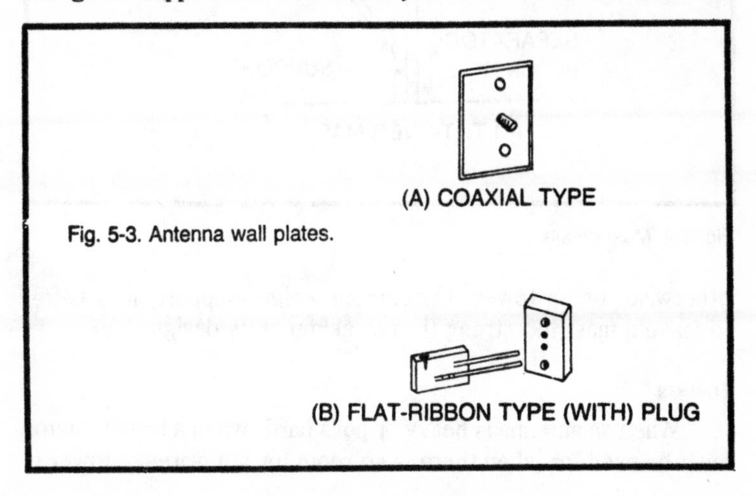

Fig. 5-3. Antenna wall plates.

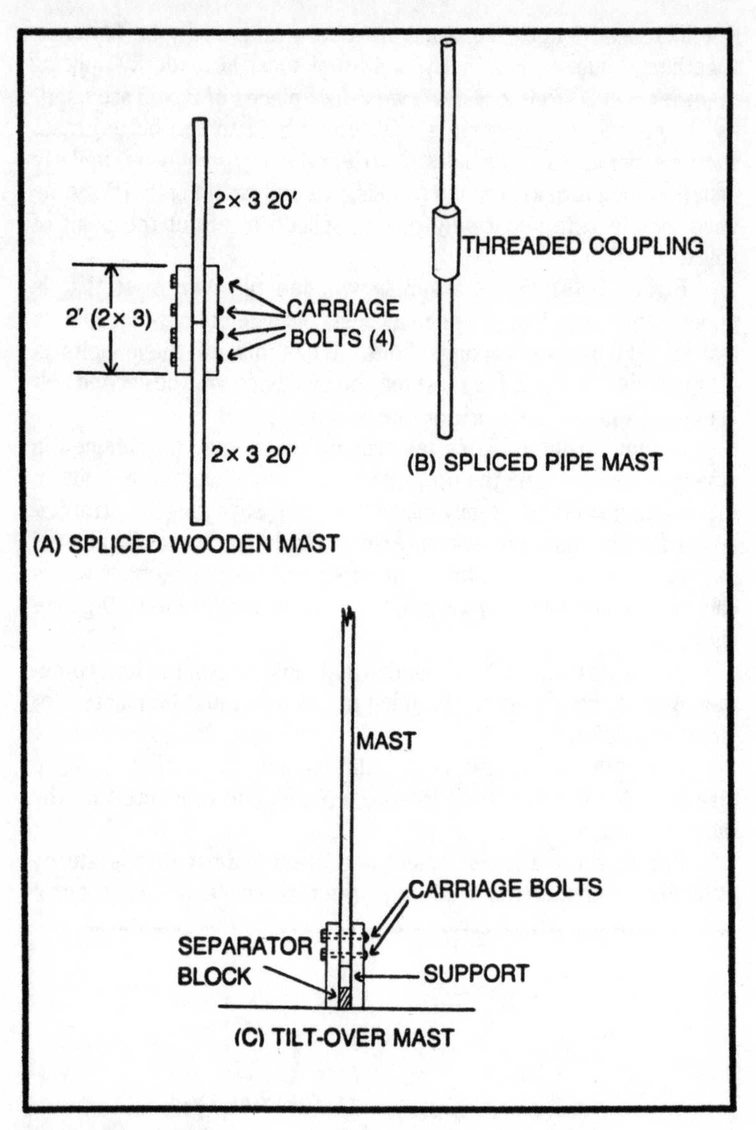

Fig. 5-4. Mast details.

otherwise, use a tower. Concerning weight support, it is better to make a mast too strong than to skimp in its design.

Towers

When an antenna is heavy or pulls hard, when a heavy rotator must be used, or when there is no room for guy wires, a tower is preferable to a mast. (Some towers also must be guyed, however.) A solidly made and correctly installed tower gives permanence and stability. Moreover, it can be climbed with a great deal more assurance than most masts can be scaled. Figure 5-5 shows the basic construction of the familar triangular tower. For good weathering and fire protection, a tower should be made of metal.

Considerable structural engineering know-how and much hard work are needed to produce a satisfactory and safe tower. This matter is beyond the scope of this book; we feel that the antenna user who needs a tower is best advised to buy one.

A metal tower must be grounded for lightning protection.

Chimney

A chimney often serves as the support for an antenna, at either or both ends (see Fig. 5-6A). The user who elects this method of support should determine first that the chimney is solid enough that the antenna will not pull it down.

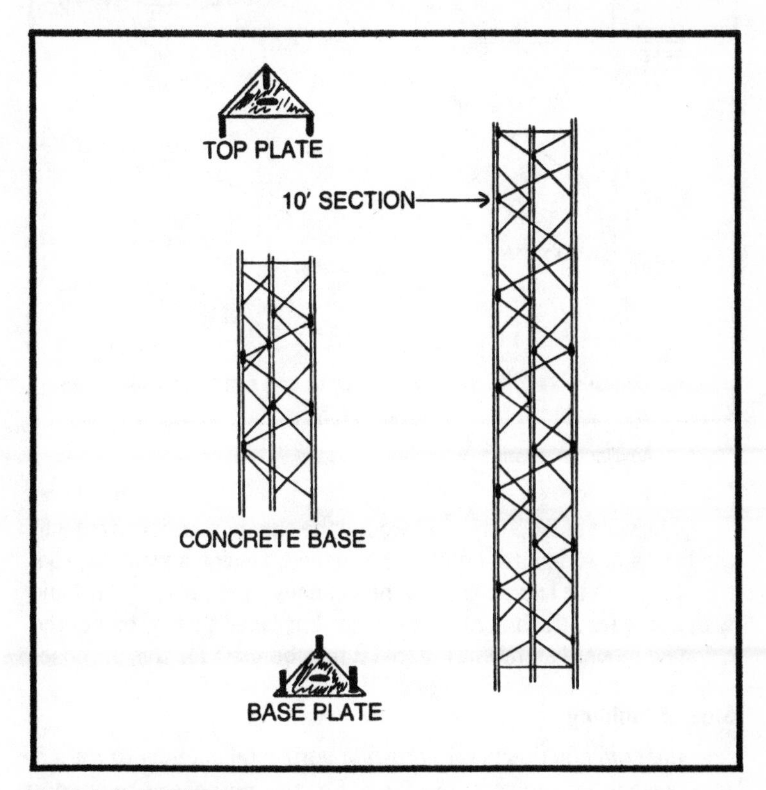

Fig. 5-5. Detail of tower.

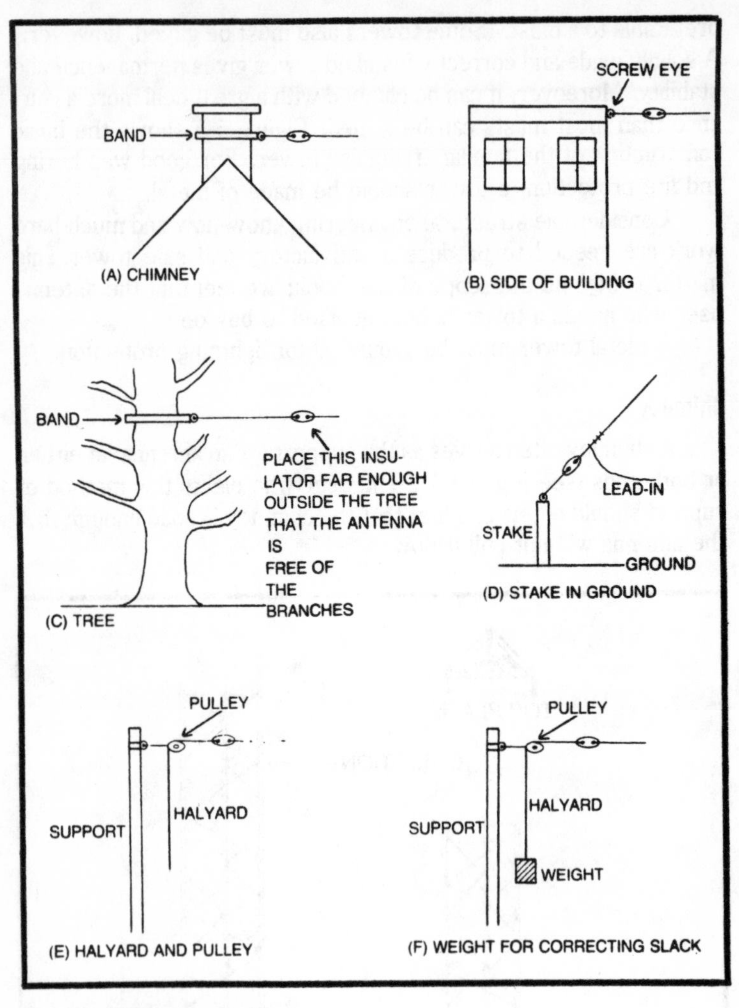

Fig. 5-6. Auxiliary supports.

The wire must not be tied around the chimney, as it can cut into the bricks or slide into the mortar. Instead, fashion a weatherproof metal band to fit snugly around the chimney, and provide this band with a ring for attachment of the wire. For labor saving, one of the chimney mounts shown in Fig. 5-7B may be used for this purpose.

Side of Building

It is common practice to attach a horizontal antenna to the side of a building, as shown in Fig. 5-6B. For this purpose, a heavy-duty galvanized screw eye is used as shown. A heavy nail and wire hook also might be employed. After the screw eye has been set, the area around its entry should be properly sealed to prevent leaks. Hook the antenna as close as possible to the top of the building.

Tree

A tree makes a handy and inexpensive support for one end of a horizontal antenna (see Fig. 5-6C). Not every tree is suitable, however. The tree should be large and steady in the wind. A large, stout branch is better than a small one, since the larger one is less subject to bending. The trunk of the tree is usually the best place to hook the antenna. For this purpose, do not drive a nail, spike, or screw into the tree, but make a band which will grip the trunk without bruising. Find out from your local tree surgeon if the band should be metal or nonmetal for the particualr tree you plan to use.

Stake in Ground

The lower end of an inclined antenna is fastened to some supporting member close to the ground. A stake often is driven into the ground for this purpose (see Fig. 5-6D). Other supports include posts, fences, shacks, and so on. In some inclined antennas, the lead-in or feeder is at the bottom, as in Fig. 5-6D; in others, it is at the top, as in Figs. 2-29A and 2-30 in Chapter 2.

Pulleys and Halyards

While for simplicity the antenna is shown in Fig. 5-6 to be attached directly to the support, it more often is fastened to a halyard run through a pulley for easily hoisting and lowering it in installation, inspection, and repair. This arrangement is shown in Fig. 5-6E. The pully must be weatherproof and corrosion resistant, and the halyard must be stout, though flexible. Rope may be employed as a halyard, but is subject to rotting and fraying, and needs to be replaced often. Flexible metal cable is satisfactory if it is rustproof. Fiberglass guy wire makes a good halyard when a nonmetal other than rope is preferred.

After it is pulled tight, the lower end of a halyard may be tied to a screw eye or hook on the mast. Alternatively, a heavy weight may be fastened to the lower end of the halyard (see Fig. 5-6F) and will automatically keep the halyard pulled tight.

Special Pointers

1. When an antenna is supported above a rooftop, it must be high enough to walk under to prevent its being a hazard to workmen and fire-fighters.

2. Do not fasten a private antenna to a utility pole or electric tower under any circumstances.

3. Do not fasten a private antenna to a radio station tower without permission from the station authorities.

4. Do not fasten an antenna to a chimney or any other part of a building without permission from the owner or other person in control of the building.

5. Do not run an antenna across a street, highway, road, alley, or public driveway.

6. Unless it is unavoidable and there are no regulations against it, do not install a mast or tower where it might fall across a power line or telephone line.

MAST MOUNTS

A fixed-station antenna stands on the ground or on a building. There are various ways of mounting the mast. If it stands on the ground, for example, its end may be driven deep enough to give good support, or it may be set in concrete, as shown in Fig. 5-7A. If the mast is attached to a building, it could be fastened to the roof, lashed to a chimney or vent pipe with cable or rope, or nailed to the side of the building. But an assortment of special hardware is commercially available which simplifies and professionalizes the job of mounting the mast. Some of the many available fixtures of this kind are shown in Figs. 5-7B through 5-7E. While these are basically TV antenna hardware, they are strong enough for amateur, CB, and SWL antennas that are not excessively tall and heavy.

Figure 5-7B shows two chimney mounts. The first (Calectro U2-000) holds the mast 4 inches from one corner of the chimney; the second (Calectro U2-010) holds the mast 3 inches from one side of the chimney. The corner mounting accommodates masts up to 1.5 inches in diameter, and the side mounting accommodates masts up to 1.75 inches in diameter. Both are easily assembled and are safe for the chimney.

Figure 5-7C shows a roof mount. The main bracket of this fixture (Calectro U2-008) is fastened to the roof by means of three screws, following which the holes must be properly sealed to prevent leaks. The mast can be inserted while it is lying down on the roof, then swung vertical and the swivel tightened. This allows guy

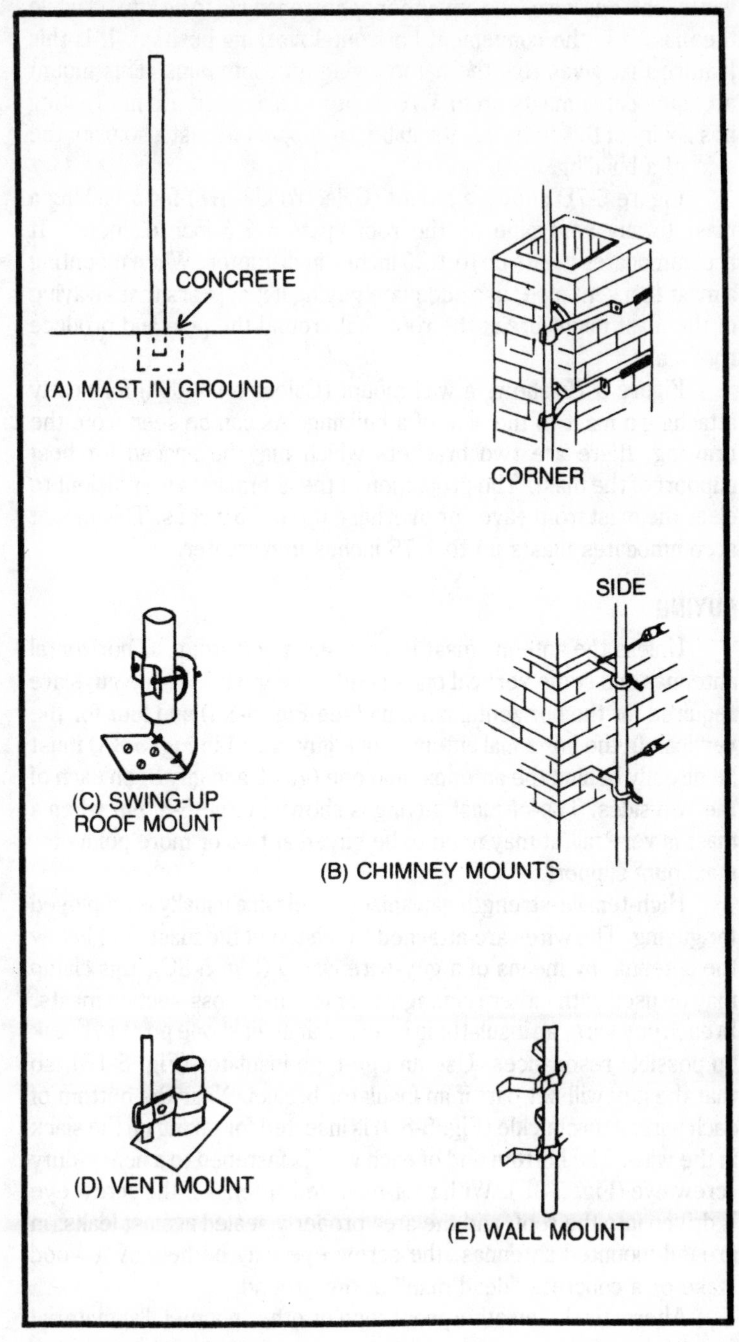

Fig. 5-7. Mounting of mast.

wires, antenna wire, and any other appurtenances to be affixed while the mast is in the convenient horizontal working position. It is this feature that gives this the name swing-up roofmount. This mount accommodates masts up to 1.75 inches in diameter. An interesting possibility of this fixture is its ability to support a mast also from the side of a building.

Figure 5-7D shows a mount (Calectro U2-018) for attaching a mast to any vent pipe on the roof up to a 2.5-inch diameter. It accommodates masts up to 1.25 inches in diameter. When mounting a mast to a vent pipe, use adequate guying if it appears that swaying of the mast might break the roof seal around the pipe and produce roof leaks.

Figure 5-7E shows a wall mount (Calectro U2-002) for neatly attaching a mast to the side of a building. As can be seen from the drawing, there are two brackets which may be spaced for best support of the mast. The projection of these brackets is sufficient to clear the mast from eaves or overhang up to 2.5 inches. This mount accommodates masts up to 1.75 inches in diameter.

GUYING

Unless the antenna mast is very short and stout, all horizontal antennas and some vertical ones need to be guyed. Three guys are required for the horizontal antenna (see Fig. 5-8A) and four for the vertical. In the horizonal antenna, one guy (guy 1 in Fig. 5-8A) must be directly behind the antenna, and one (guy 2 and guy 3) on each of the two sides. Top-of-mast guying is shown in Fig. 5-8; but when a mast is very tall, it may need to be guyed at two or more points for maximum support.

High-tensile-strength galvanized steel wire usually is employed for guying. The wires are attached to the top of the mast, just below the antenna, by means of a guy-wire clamp (Fig. 5-8C); this clamp may be used with either rectangular or circular cross-section masts. In each guy wire, an insulator is inserted at at least one point to break up possible resonances. Use an egg-type insulator (Fig. 5-1B), so that the guy will not part if an insulator breaks. Near the bottom of each wire, a turnbuckle (Fig. 5-8D) is inserted for taking up the slack in the wire. The bottom end of each wire is fastened to a heavy-duty screw eye (Fig. 5-8E). With roof-mounted antennas, the screw eye is driven into the roof, and the area properly sealed against leaks; in ground-mounted antennas, the screw eye may be held by a wood stake or a concrete "dead man" in the ground.

Alternatively, weatherproof rope or other nonmetallic material may be used for guys, especially in obstinate cases of absorption or

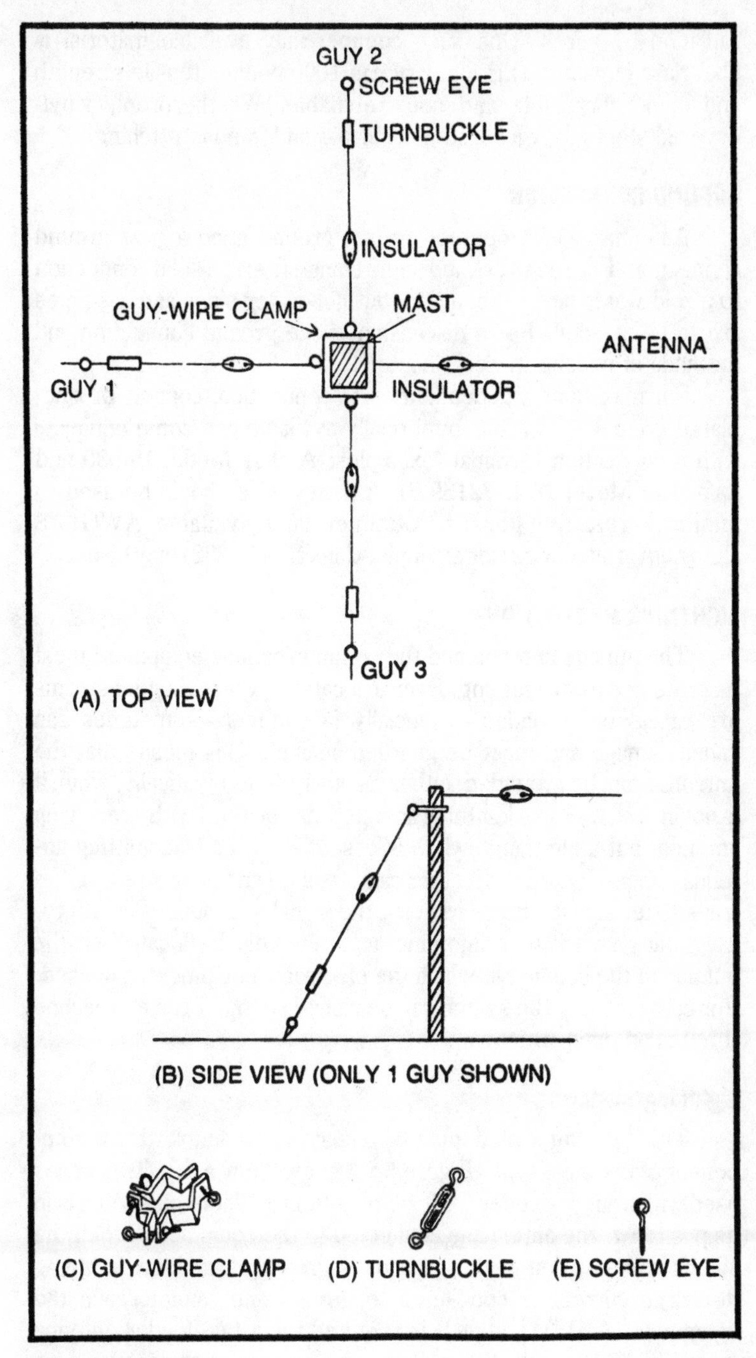
GUY 2
SCREW EYE
TURNBUCKLE
INSULATOR
GUY-WIRE CLAMP
MAST
ANTENNA
GUY 1
INSULATOR
GUY 3
(A) TOP VIEW
(B) SIDE VIEW (ONLY 1 GUY SHOWN)
(C) GUY-WIRE CLAMP
(D) TURNBUCKLE
(E) SCREW EYE

Fig. 5-8. Guying.

radiation by wires. One such commercially available material is fiberglass guy line. This material has 1000 pounds tensile strength and is nonshrinkable and nonstretchable. Weatherproof, vinyl-covered steel guy wire also is available and is nonstretching.

GROUND CONNECTION

Antennas which operate against ground need a *good* ground connection. For receivers and some transmitters, a solid connection to a cold-water pipe is adequate, but hot-water pipes and gas pipes are to be avoided. For a discussion of the ground connection and methods of making it, see Chapter 2.

For making an independent earth connection, copper- or zinc-plated ground rods are commercially available and come equipped with a connection terminal (examples: Archer Model 15-530 and Lafayette Model 18 R 72159W). A heavy wire should be used to minimize resistive losses. Commercially available AWG #8 aluminum ground wire (for example, Calectro U4-332) may be used.

LIGHTNING PROTECTION

The outside antenna and the communications equipment must be protected from lightning. Even in locations where thunder storms are rare, static buildup—especially in summer—sometimes can cause damage and must be guarded against. This means that the antenna must be grounded, either manually or automatically, when it is not in use. A simple lightning arrester may be used with a receiving antenna, but a lightning switch is essential with a transmitting antenna, since a lightning arrester can break down under stress of the transmitter signal, short-circuiting the signal to ground. For safety, a lightning switch *or* a lightning arrester must be located on the outside of the building in which the electronic equipment is housed. For convenience, the switch can be placed so that it can be reached through an open window.

Lightning Switch

The lightning switch must be a heavy-duty double-throw component of the knife type. Figure 5-9A shows how a SPDT switch is used with a single-feeder (or lead-in) antenna. When this switch is in its positon *a*, the antenna is connected to the equipment; when the switch is in position *b*, the antenna is grounded, and any lightning discharge current is conducted to ground and cannot reach the equipment. A DPDT switch is required for a two-feeder antenna (see Fig. 5-9B). With the switch in its positon *a*, the feeders are connected to the equipment; when the switch is in position *b*, the two halves of the antenna are both grounded, and any lightning discharge current is conducted to ground and cannot reach the equipment. A heavy ground wire must be used with the switch. An example is commercially available aluminum ground wire, such as Calectro U4-353.

Lightning Arrester

A lightning arrester is a small spark gap connected between antenna and ground. Being an open circuit, it cannot normally interfere with the antenna signal. A lightning discharge produces high voltage, and the arrester automatically flashes over and conducts the resulting current to ground, around the equipment. Figure 5-9C shows a single-gap arrester used with a single-feeder (or lead-in) antenna. Figure 5-9D shows a double-gap arrester used with a two-feeder antenna. A heavy ground wire must be used with the arrester, such as Calectro UA-353.

Lightning arresters are available in several makes and models. One type is provided with simple terminal screws. The wires must be stripped and passed under the screw heads. Another type has terminals which when tightened bite through the insulation of the wires, thus requiring no stripping.

Protection of Control Wires

Power and control leads to an antenna rotator, as well as antenna leads, must be protected from lightning. At least one arrester (JFD Model AT104S) accommodates four-wire rotator cable, and one other arrester (JFD Model AT106S) accommodates five-wire rotator cable.

VEHICULAR ANTENNAS

Vehicular antennas (car, truck, and boat) are nearly always vertical. They are unguyed and base supported. Almost all are factory built.

The chief concerns of the installer of these antennas are that the base insulation is adequate, especially where a transmission line passes through the metal body of the vehicle, and that the installation be mechanically solid. For mechanical details of CB mobile antenna installation, see Chapter 3.

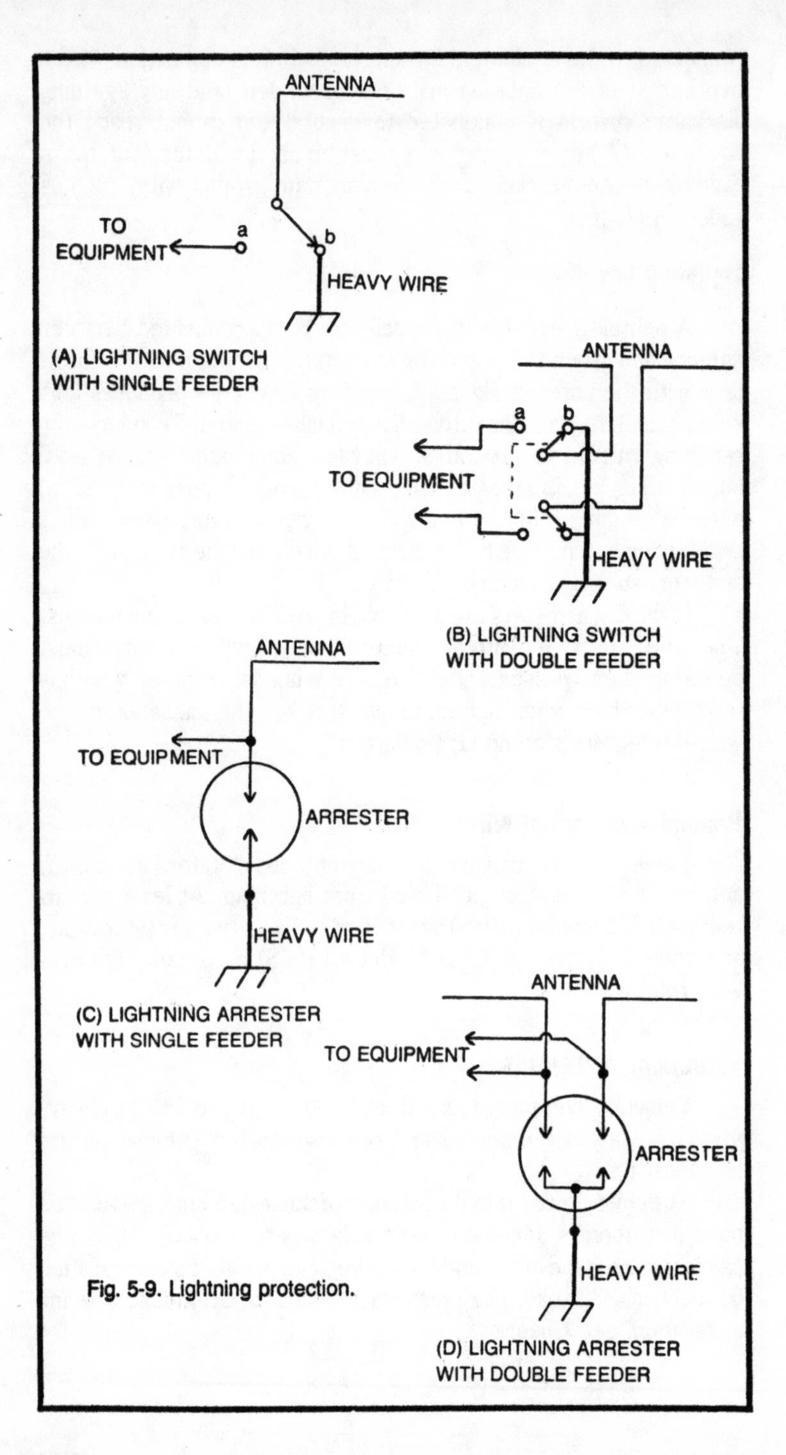
ANTENNA
TO EQUIPMENT
a
b
HEAVY WIRE
(A) LIGHTNING SWITCH WITH SINGLE FEEDER
ANTENNA
a
b
TO EQUIPMENT
HEAVY WIRE
(B) LIGHTNING SWITCH WITH DOUBLE FEEDER
ANTENNA
TO EQUIPMENT
ARRESTER
HEAVY WIRE
(C) LIGHTNING ARRESTER WITH SINGLE FEEDER
ANTENNA
TO EQUIPMENT
ARRESTER
HEAVY WIRE
(D) LIGHTNING ARRESTER WITH DOUBLE FEEDER
Fig. 5-9. Lightning protection.

LAWS AND SAFETY RULES IN ANTENNA INSTALLATION

Before finalizing plans for an antenna, study carefully any municipal and federal regulations that pertain to antenna installation, and abide by them. Also, study the current electrical code as it may apply to antennas, lead-ins, and rotator cables. Become familiar with local environmental impact rules applying to towers and masts.

For regulations regarding the height of CB antennas, see Figs. 3-4 and 3-5 in Chapter 3.

Chapter 6

Tests & Measurements

The rough-and-tumble appraisal of a transmitting antenna in terms of casual signal reports from listeners, or of a receiving antenna on the basis of audible strength of received signals, is unreliable, to say the least. Abundant procedures, as well as numerous instruments at all price levels, are available for testing antenna performance. Every antenna—especially the transmitting type—should be carefully tested before it is put into permanent service and should be checked periodically to insure that its performance remains satisfactory.

This chapter describes in simple terms the principal tests for antennas. In some instances, instructions are given for performing the same class of test with different instruments or devices, one instrument usually being cheaper or more widespread than the others. In many cases, specialized test instruments are not required; instead, a general-purpose instrument is used.

CAUTION. Many antenna tests require that an RF test signal be applied to the antenna, and this signal—no matter how weak—is radiated by the antenna. The test therefore can cause interference. Keeping this in mind, the operator must use the minimum signal that will insure a satisfactory measurement, keep a test as short as possible, and identify his own station (if he is licensed) as often during the test as is required by FCC rules.

SIMPLE STANDING-WAVE DETECTION

Several simple methods allow standing waves to be detected on an antenna wire or feeder wire. The devices used indicate either current loops and nodes or voltage loops and nodes as noted below.

Incandescent Bulb

A small, filament-type pilot lamp with a pickup loop (see Fig. 6-1A) makes an inexpensive detector. This is a current-indicating detector. The lamp is a 2-volt or 6-volt type (either a screw-base or bayonet-base type can be used). The pickup loop is a 1-turn coil of solid, insulated hookup wire wound 1 inch in diameter and soldered directly to the base contacts of the lamp.

Test Procedure. (1) Switch the transmitter on. (2) Hold the glass envelope of the bulb and position the loop close to, but not touching, the antenna or feeder, as shown in Fig. 6-1A. (3) While maintaining constant spacing between the loop and wire, move the loop slowly along the wire. (4) If the brightness of the bulb does not change, no standing waves are present. (5) If standing waves are present, the bulb will brighten at each loop point and will dim (or extinguish) at each node point.

Neon Bulb

A small neon bulb, like the filament-type lamp, also makes a simple, inexpensive detector (see Fig. 6-1B). Unlike the incandescent lamp, the neon bulb requires no pickup coil. It is, however, a *voltage*-indicating detector. The neon bulb may be a quarter-watt type; but with some strong transmitters, the 1- or 2-watt type may be required.

Test Procedure. (1) Switch the transmitter on. (2) Hold the glass envelope of the bulb and touch the metal base of the bulb to the antenna or feeder, as shown in Fig. 6-1B. (3) While maintaining constant pressure of the bulb against the wire, slide the bulb slowly along the wire. (4) If the brightness of the bulb does not change, no standing waves are present. (5) If standing waves are present, the bulb will brighten at each loop point and will dim (or extinguish) at each node point.

Thermocouple-Type Meter

A radio-frequency meter of the thermocouple type (RF ammeter, RF milliammeter, or thermogalvanometer) becomes a standing-wave detector when a small pickup loop is connected directly to its two terminal screws (see Fig. 6-1C). This is a *current*-indicating detector. The pickup loop is a 1-turn coil of solid, insulated hookup wire wound 1 inch in diameter. Use a thermogalvanometer or RF milliameter (e.g., 0–500 mA) for low-powered transmitters, an RF ammeter (e.g., 0–1A) for high power.

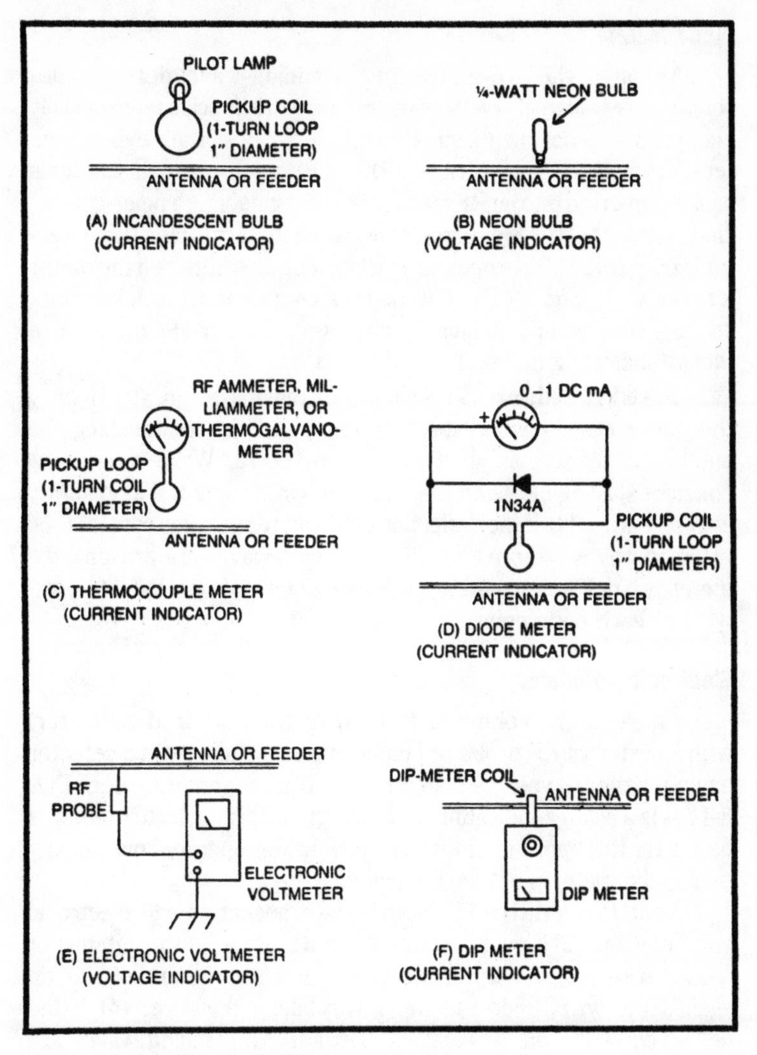

Fig. 6-1. Standing-wave detectors.

Test Procedure. (1) Switch the transmitter on. (2) Holding the meter case, position the loop close to but not touching the antenna or feeder, as shown in Fig. 6-1C. (3) While maintaining constant spacing between the loop and wire, move the loop slowly along the wire. (4) If the deflection of the meter remains constant, no standing waves are present. (5) If standing waves are present, the meter will deflect upward at each loop point and will deflect downward at each node point.

Diode Meter

Although the RF-meter-type standing-wave detector described previously is very simple, many technicians—especially hobbyists—do not own such a meter. Most, if not all, experimenters, however, do possess a 0–1 DC milliammeter, and this meter is easily converted into an RF instrument by connecting a point-contact diode (type 1N34A, for example) between its terminals. The detector is completed by connecting a pickup loop also between the meter terminals (see Fig. 6-1D). The loop is a 1-turn coil of solid, insulated hookup wire wound 1 inch in diameter. This diode meter is a *current*-indicating detector.

Test Procedure. (1) Switch the transmitter on. (2) Holding the meter case, positon the loop close to, but not touching the antenna or feeder, as shown in Fig. 6-1D. (3) While maintaining constant spacing between the loop and wire, move the loop slowly along the wire. (4) If the deflection of the meter remains constant, no standing waves are present. (5) If standing waves are present, the meter will deflect upward at each loop point and will deflect downward at each node point.

Electronic Voltmeter

An electronic voltmeter (VTVM or transistorized voltmeter) with an external RF probe will function as a standing-wave detector having several voltage ranges. This arrangement—see Fig. 6-1E—is a *voltage*-indicating detector. A battery-operated meter is better for this application than is a power-line-operated one. In any event, the meter must be grounded.

Test Procedure. (1) Switch the transmitter and electronic voltmeter on. (2) Touch the top of the RF probe to the antenna or feeder wire. (3) While maintaining constant pressure of the tip against the wire, slide the top slowly along the wire. (4) If the deflection of the meter remains constant, no standing waves are present. (5) If standing waves are present, the meter will deflect upward at each loop point and will deflect downward at each node point.

Dip Meter

By the simple flip of a switch, most dip meters (grid-dip meter, FET dipper, etc.) are transformed into a simple absorption wavemeter with meter readout; the internal power supply is automatically switched off, so the AC-type instrument need not even be plugged into the power line. As a wavemeter, the dip meter is an excellent standing-wave detector; in this function, it is a tuned, ***current*-**indicating detector.

Test Procedure. (1) Set the dip-meter (DM) function switch to its OSCILLATOR-OFF position. (2) Switch the transmitter on. (3) Tune the DM to the transmitter frequency by coupling the DM temporarily to the transmitter and tuning the DM for peak deflection of the meter. Do not subsequently disturb the setting of the tuning dial. (4) Couple the DM to the antenna or feeder by holding the DM coil close to the wire, with the turns of the coil parallel to the wire. (5) While maintaining constant spacing between the coil and the wire, move the DM along the wire. (6) If the deflection of the meter undergoes no change, no standing waves are present. (7) If standing waves are present, the meter will deflect upward at each loop point, and will deflect downward at each node point.

CHECKING ANTENNA OR FEEDER CURRENT

Occasionally, especially in the design and development of a new antenna, it is important to measure the current in an antenna or feeder by temporarily cutting the wire and inserting a radio-frequency ammeter. When this is necessary, the technician must know whether or not standing waves are present and what their distribution is. For this purpose, he may use one of the techniques just described.

If there are no standing waves, the current has the same value at all points along the wire, and the meter may be inserted at the most convenient point. But if there are standing waves, the meter must be inserted at a known current loop in order to obtain maximum current reading. Figure 6-2 shows how an ammeter will give different readings at different points of insertion in a full-wave-long wire having standing waves. At *B* and *D*, the current loop causes the meter to read maximum. At *A*, the meter reads low because the loop is not yet fully developed; and at *C*, the meter reads zero (or a very low minimum) because here we have a current node.

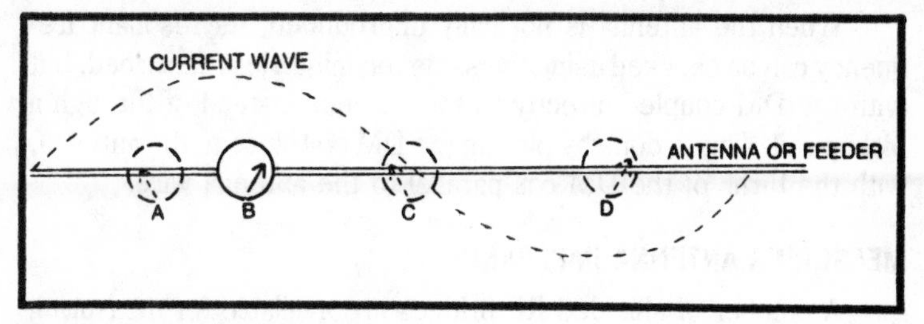

Fig. 6-2. Variation of RF current.

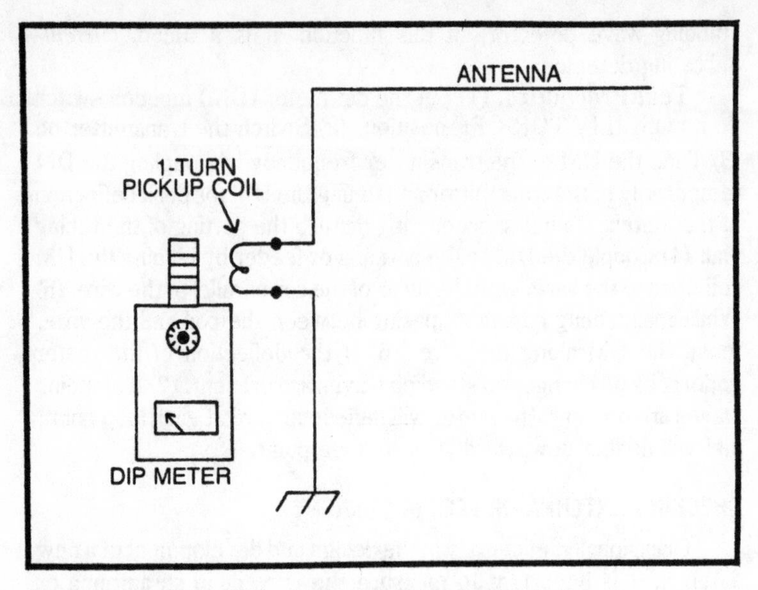

Fig. 6-3. Using dip meter to check resonant frequency of antenna.

CHECKING RESONANT FREQUENCY OF ANTENNA

The resonant frequency of a simple, grounded antenna can be checked with the setup shown in Fig. 6-3. Here, a dip meter (DM) is inductively coupled to the antenna through a 1-turn coil (insulated hookup wire tied around the DM coil) connected in the lead-in or single-wire feeder between antenna and ground.

Test Procedure. (1) Switch on the DM. (2) Tune the DM *from its highest frequency downward*, changing coils as required, and watching closely for dips. (2) Usually, a dip will be obtained at the resonant frequency and also at one or more harmonics. The fundamental resonant frequency is the lowest frequency at which dip occurs, hence the importance of tuning *downward* throughout the DM range.

When the antenna is normally ungrounded, its resonant frequency can be checked using the same technique just described, but with the DM coupled directly to the antenna instead of through a pickup coil. This is done by placing the DM coil close to the antenna, with the turns of the DM coil parallel to the antenna wire.

MEASURING ANTENNA IMPEDANCE

A number of shielded RF bridges are available for measuring antenna impedance. These instruments (sometimes called impedance meters) span the quality grades from laboratory-type to hobbyist-type, and some in the latter category are available in kit form. This impedance bridge requires an RF input signal at the desired test frequency; and for this purpose, the instrument is connected between a transmitter and the antenna or between a suitable RF signal generator and the antenna. In many cases, this means simply inserting the bridge in the regular transmission line between transmitter and antenna.

Test Procedure. (1) Place the bridge as shown in either of the setups in Fig. 6-4. (2) Switch on the transmitter or signal generator. (3) Set the balance dial of the bridge for null. (4) At null, read the impedance from the dial.

Common bridge input impedance is 50 ohms. This permits insertion into the station coaxial transmission line (Fig. 6-4A) or between the low-impedance output of a signal generator and the coaxial line (Fig. 6-4B). Some bridges have two output binding posts, in addition to coaxial fixtures, to permit connection of a 2-wire feeder or single-wire feeder and ground.

CHECKING TRANSMISSION-LINE IMPEDANCE

The impedance of a transmission line can be measured with an impedance bridge. Use of an RF impedance bridge to measure

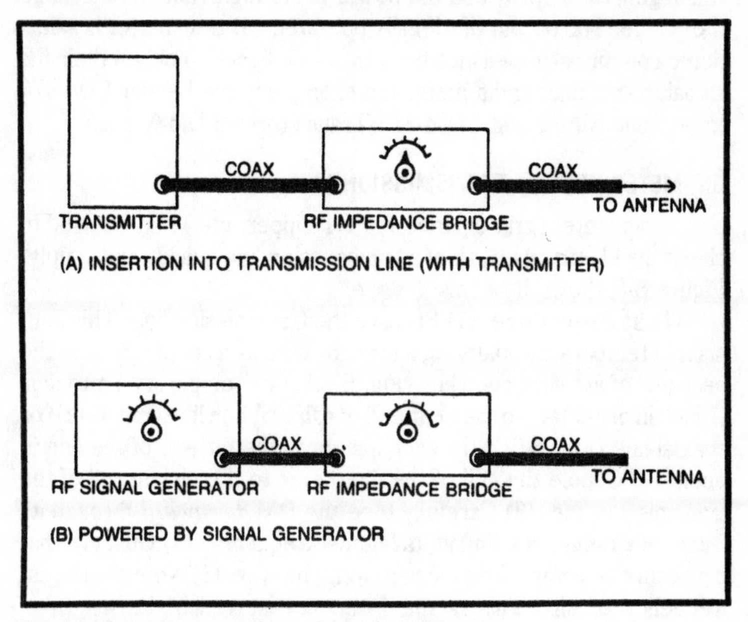

Fig. 6-4. Setup for measuring antenna impedance.

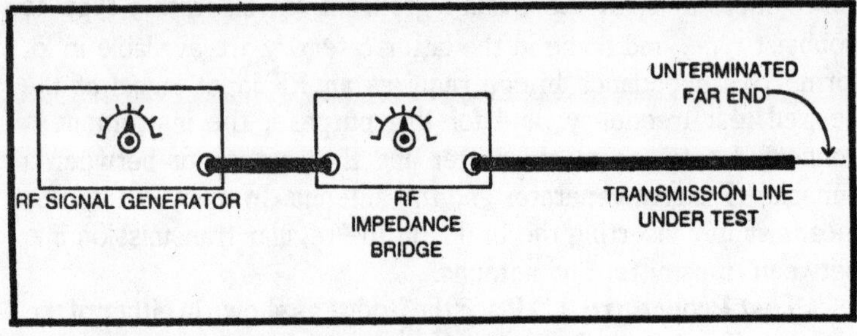

Fig. 6-5. Measuring impedance of transmission line.

antenna impedance has already been described. The same technique is employed to check a transmission line, except that the line, instead of the antenna, is connected to the terminals (or coaxial jack) of the bridge (see Fig. 6-5).

Test Procedure. (1) Following Fig. 6-6, connect a suitable RF signal generator (having up to 1-volt output) to the input jack or terminals of the bridge. (2) Connect the transmission line to the *unknown* jack or terminals of the bridge. The far end of the line must be open circuited. (3) Switch on the signal generator. (4) Adjust the bridge for null and at that point read the unknown impedance from the main dial of the bridge. (A transmitter may be used in lieu of the signal generator, provided the power is not high enough to damage the bridge. The output of a legally operated CB transmitter is safe.) Some commrcial impedance meters are designed to obtain their RF signal from a mating dip meter (an example is the Leader LIM-870 Impedance Meter and LDM-815 Transistorized Dip Meter).

DIP-METER TEST OF TRANSMISSION LINE

A dip meter (grid-dip meter, FET dipper, etc.) may be used to check the electrical length of a transmission line quickly and simply. Figure 6-6 shows how this is done.

Test Procedure. (1) Prepare the transmission line. This consists of temporarily soldering a straight wire as a coupling coil to the near end of a 2-wire open line (Fig. 6-6A) or a temporary 1-turn coil (1 inch in diameter) to the near end of a flat-ribbon line (Fig. 6-6B) or coaxial line (Fig. 6-6C). In each instance, the far end of the line is open. (2) Couple the coil of the dip meter to the pickup coil of the transmission line. (3) Carefully tune the DM *downward* through its frequency range, starting at its highest frequency. (4) Observe that dip occurs at several frequencies along the way. (5) Note the lowest frequency of dip. This is the frequency at which the length of transmission line is a quarter-wavelength.

CHECKING ATTENUATION IN COAXIAL CABLE

The attenuation introduced by a length of coaxial cable can be checked with an unmodulated transmitter, an RF ammeter, and a dummy load. The dummy load is a noninductive resistor whose resistance equals the characteristic impedance of the type of coax under test and whose power rating is equal at least to the power output of the transmitter. The ammeter and the dummy load each is enclosed in a small metal shield box provided with coaxial fittings for

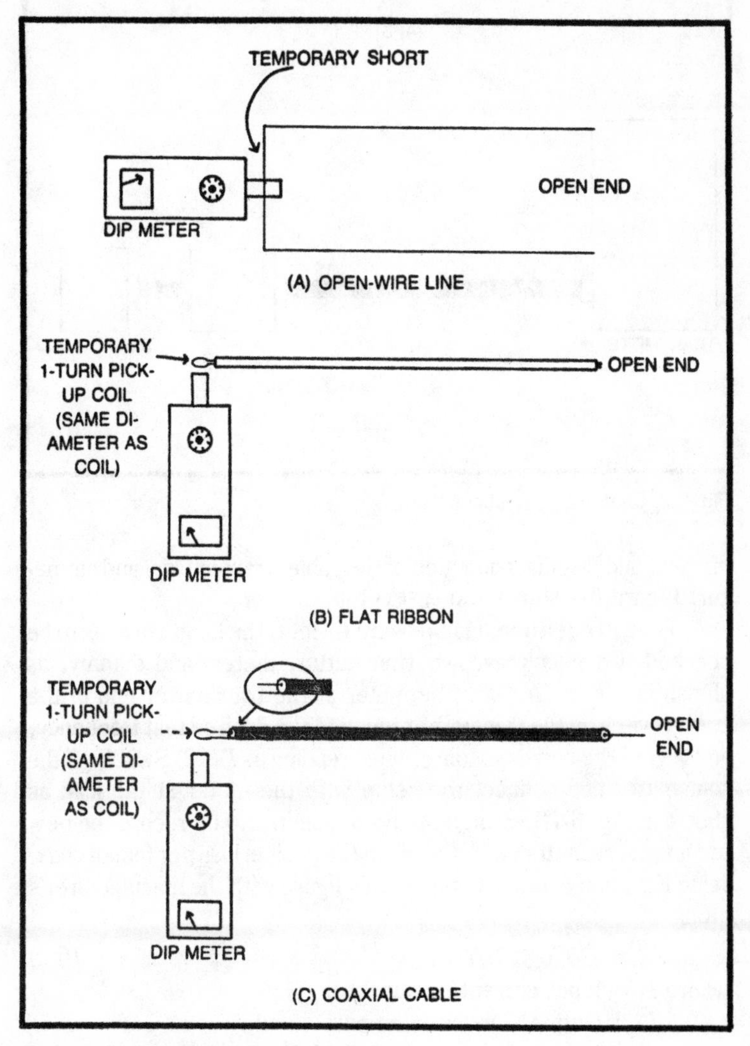

Fig. 6-6. Dip-meter test of transmission line.

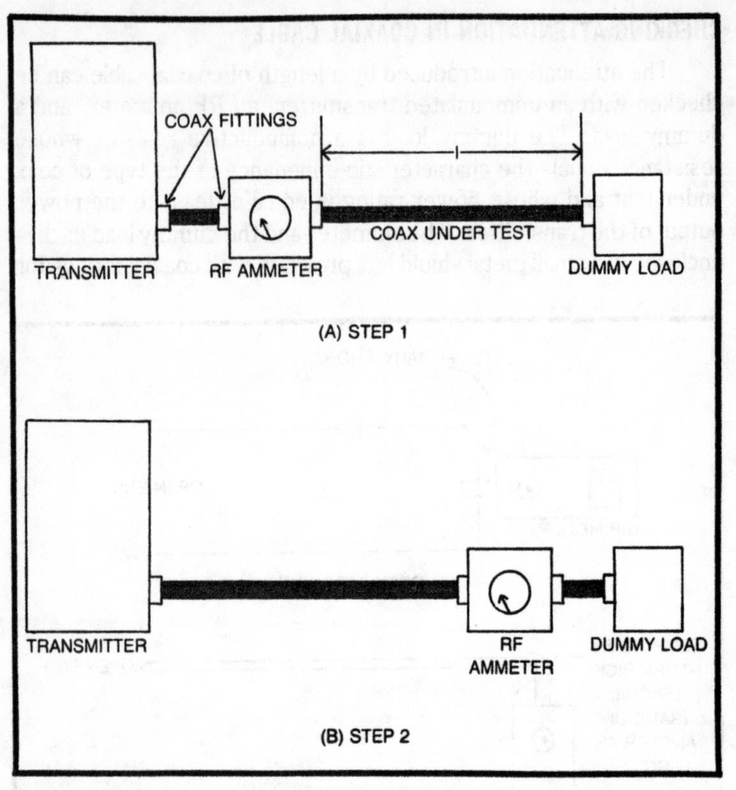

Fig. 6-7. Checking coax attenuation.

efficient and fast disconnection of the cable, transmitter, and ammeter. Figure 6-7 shows the test setup.

Test Procedure. (1) Measure (in feet) the length of coax to be checked. Connect the coax, transmitter, meter, and dummy, as shown in Fig. 6-7A, with the meter at the transmitter end of the coax. Switch on the transmitter tuned to the desired test frequency, and record the corresponding meter reading as I_1. (2) Switch off the transmitter and connect the setup with the meter at the load as shown in Fig. 6-7B. Switch on the transmitter, and record the new reading of the meter as I_2. Calculate the decibel loss per foot of coax, using Equation 6-1, and compare this figure with the manufacturer's rating.

$$\text{dB/ft} = (20 \log_{10} I_2/I_1)/l \qquad (6\text{-}1)$$

where I_1 = Input current in amperes,
I_2 = Output Current in amperes, and
l = Length of coax sample in feet.

CHECKING STANDING-WAVE RATIO

A low value of standing-wave ratio (SWR)—ideally 1.0:1—is very important for efficient operation of an antenna and for protection of a transmitter. The subject of standing-wave ratio is explained in detail in Chapter 3.

Several SWR meters are available in laboratory and hobbyist grades. Some of these are combined with RF wattmeters, and some (like CB test sets) with wattmeters and impedance meters. The most popular type of SWR meter presently in use outside of the microwave spectrum is the reflectometer type, the essentials of which are shown in Fig. 6-8A. In this type, RF energy from the transmitter (or equivalent signal generator) on its way to the antenna passes through a coaxial chamber consisting of an inner metal rod and outer metal shell. Each of two straight pickup wires mounted near the rod, picks up energy inductively from the latter and presents it to input diode D1 (for the first wire) and output diode D2 (for the second wire). The corresponding DC output of the diodes deflects microammeter M1. When switch S1 is in position 1, the meter thus is deflected by forward power; and when S1 is in position 2, M1 is deflected by reflected power. The meter scale is calibrated to read directly in standing-wave ratio.

Test Procedure. (1) Connect the SWR meter between the transmitter (or signal generator) and antenna, as indicated in Fig. 6-8B. (2) Throw switch S1 to position 1. (3) Switch on the transmitter. (4) Set potentiometer R1 for exact full-scale deflection of the meter. (5) Throw the switch to position 2. (6) Read the SWR from the meter.

If the reflectometer is homemade and no means is available for calibrating the instrument from a standard, the SWR may be calculated *approximately* from the microammeter readings: (1) With the transmitter switched-on and the antenna connected, throw switch S1 to position 1 and set potentiometer R1 for full-scale deflection of meter M1 (100 μA). (2) Throw S1 to position 2, and record the new meter reading as I (in μA). (3) SWR = $100/I$.

CHECKING FIELD STRENGTH

The performance of an antenna as a radiator is best appraised by checking the intensity of the signal emitted by the antenna. If the measurement is always made at the same location point a given distance from the station, different antennas may be compared, and the performance of a single antenna may be followed over a time period or as it is operated with different transmitters. For these

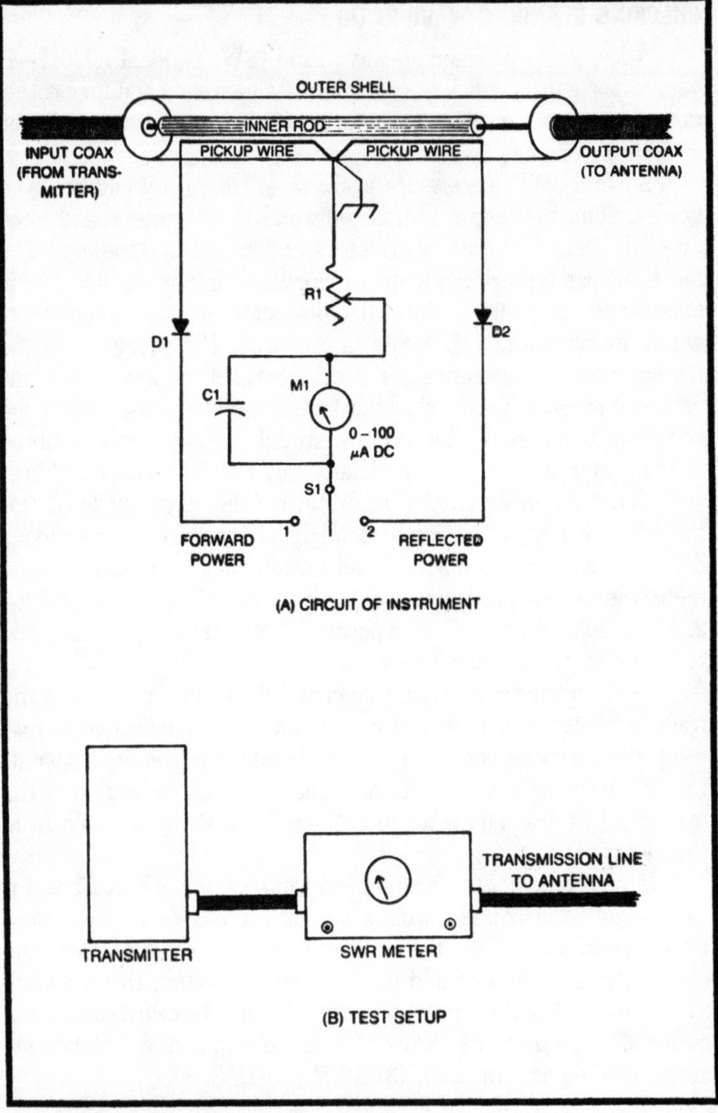

Fig. 6-8. Reflectometer-type SWR meter.

jobs, a *field-strength meter* is used, and this instrument is nothing more than a specialized portable receiver.

Field-strength meters in the engineering and service categories employ a superheterodyne circuit. A small whip antenna picks up the signal. An output meter reads the intensity of the received signal directly in microvolts and millivolts. Some CB test sets combine a field-strength meter (FSM) with an SWR meter, power meter, and impedance meter, employing one meter for all of the functions. For most amateur and hobbyist purposes, a simple tuned circuit with diode meter is sufficient for comparative measurements.

MEASURING ANTENNA DIRECTIVITY

The directional characteristic of a beam antenna can be plotted with a field-strength meter. The resulting directional pattern is drawn on polar-coordinate graph paper. Two ways of making the point-by-point measurement are shown in Fig. 6-9. Both methods require that a test signal be applied to the antenna by the regular transmitter or by a signal generator which will give a strong enough signal to actuate the field-strength meter.

In the first method, the antenna is held stationary while the field-strength meter is walked around the antenna in a *uniform* circle of at least 1 wavelength radius, and readings are taken at as many

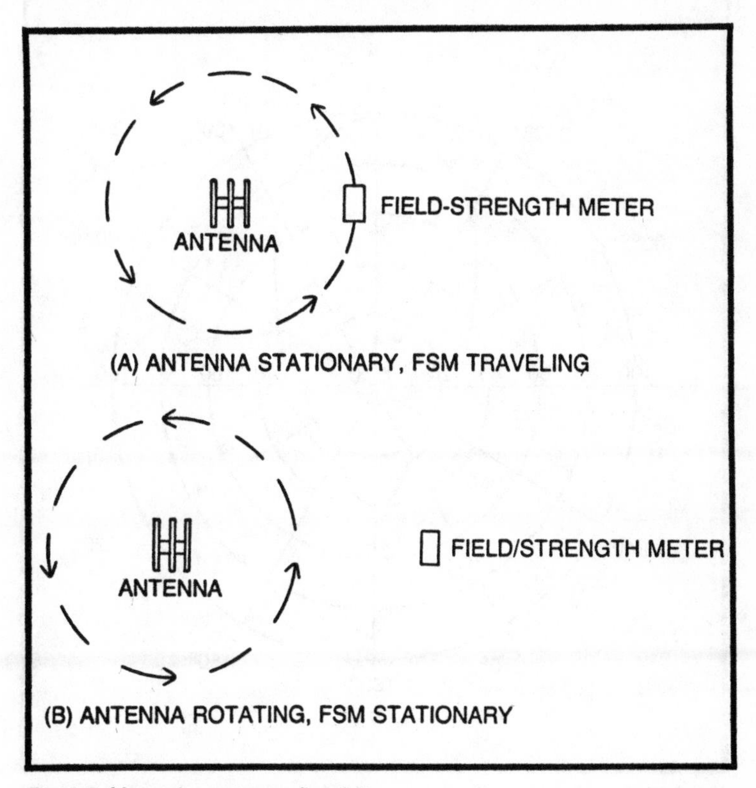

Fig. 6-9. Measuring antenna directivity.

points around the compass as possible (see Fig. 6-10). This method is difficult to pursue with reliability, since obstructions will be encountered unless the antenna and meter are in a wide-open space and unless the operator takes great pains to circle the antenna accurately.

The second method is easier. The field-strength meter is mounted stationary at a point at least 1 wavelength from the antenna, the antenna is rotated through 360 degrees, and readings are taken at as many points as possible around the circle (see Fig. 6-9B). A sample plot is shown in Fig. 6-10. Thus, the antenna is rotated a few degrees, as determined by means of a transit, protractor, or mariner's compass. At that point, a reading of field strength is taken. Then, the antenna is rotated further in the same direction, another reading taken. And so on.

Figure 6-10 shows a sample plot of directivity, i.e., field strength versus azimuth. Here, the horizontal axis can be divided

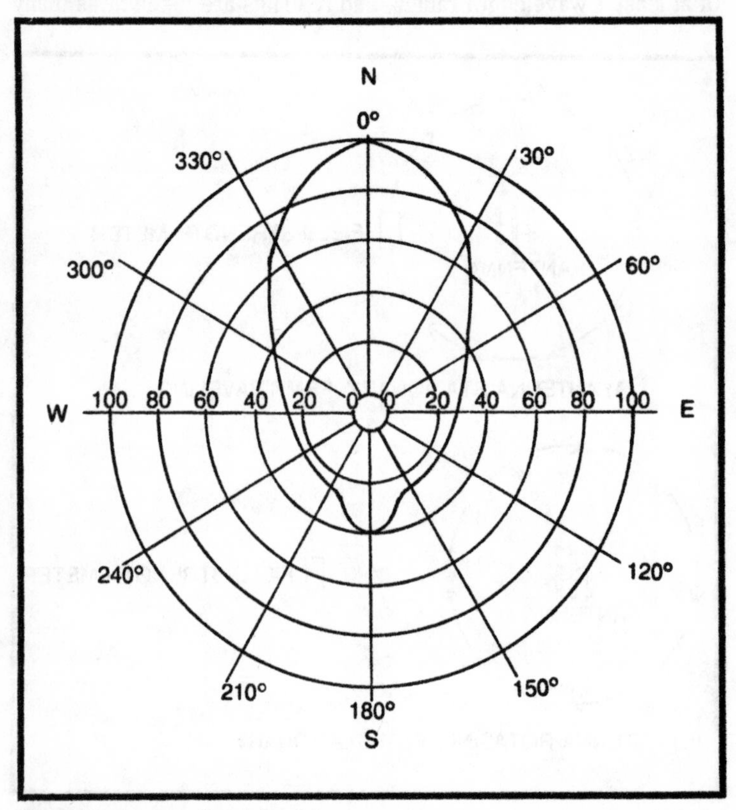

Fig. 6-10. Antenna directivity plot.

into millivolts or microvolts if a professional field-strength meter is used. For clarity and simplicity, only the major lines are shown in Fig. 6-10; standard polar-coordinate graph paper has many more lines than are shown here.

For simple determination of the front-to-back ratio of a beam antenna, only two measurements are needed: one directly in front of the antenna, and the other directly behind the antenna.

Index

Index

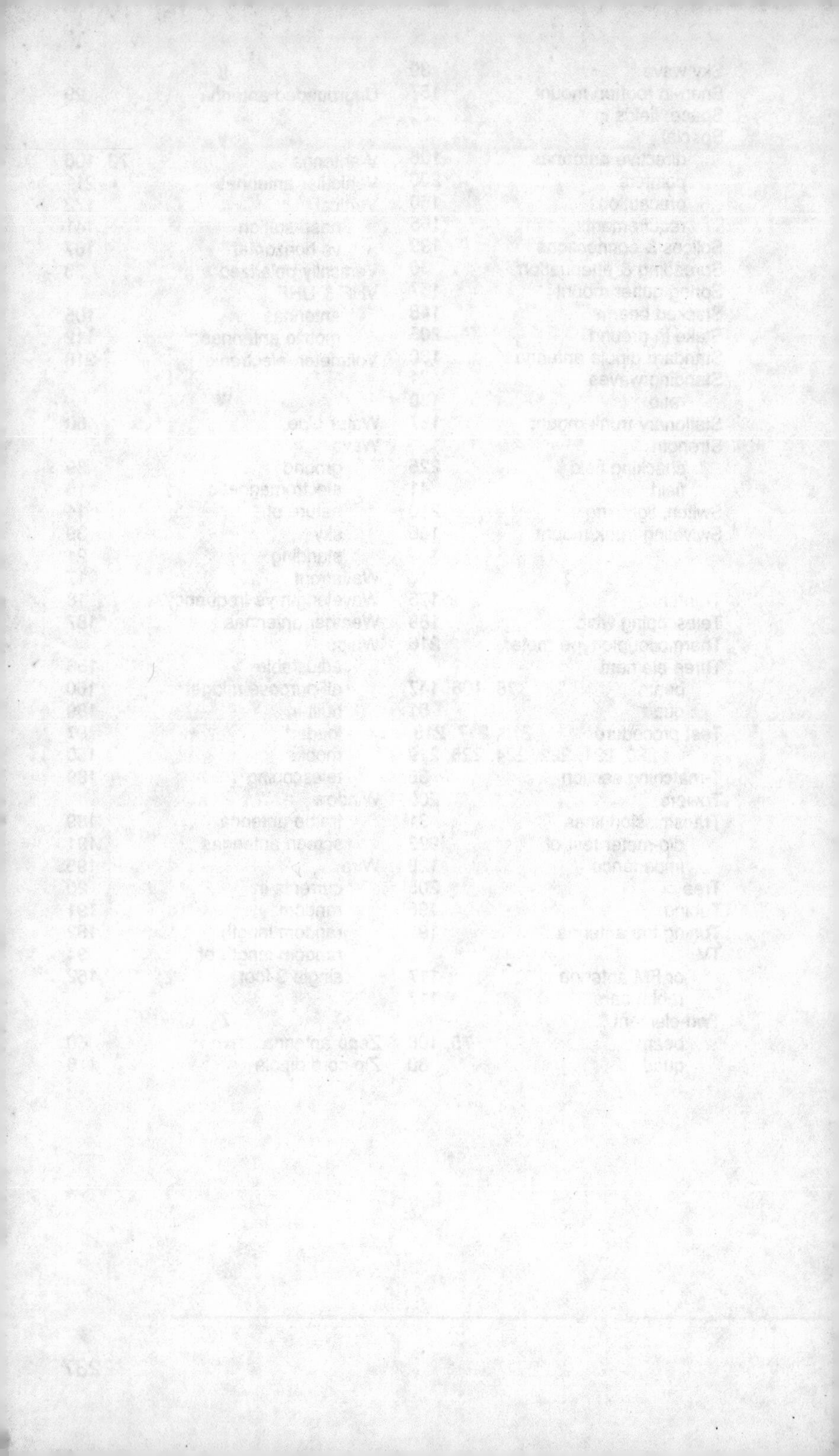